Embodied AI Education

Unlocking Human Potential Through Enactive Learning

Johnna Haskell

Published by: Epic Leaf Innovations LLC

For permission requests, please contact the publisher at: epicleafinnovations.com

Hardcover: 978-1-964562-03-2

Paperback: 978-1-964562-02-5

Table of Contents

Introduction

Picture this: a classroom where virtual reality transports students to ancient Rome, where AI tutors provide personalized feedback in real time, and where complex scientific concepts are brought to life through interactive holograms. This isn't science fiction—it's the emerging reality of education in the age of artificial intelligence.

Welcome to the frontier of learning, where the boundaries between the physical and digital worlds blur and where the possibilities for education are limited only by our imagination. If you're holding this book, chances are you're curious about how AI is reshaping the educational landscape. Maybe you're an educator looking to stay ahead of the curve, a parent wondering what your child's future classroom might look like, or simply someone fascinated by the intersection of technology and learning. Whatever brought you here, get ready for a journey that will challenge your preconceptions and open your eyes to the transformative potential of AI in education.

AI in Education: More Than Just Smart Machines

Before we learn more about this educational revolution, let's take a moment to demystify AI. At its core, artificial intelligence is about creating systems that can perceive, learn, reason, and act. In the context of education, AI isn't just about robots teaching classes (though that's not entirely off the table). It's about leveraging advanced algorithms and machine learning to enhance every aspect of the learning process.

AI in education comes in many forms. There are intelligent tutoring systems that adapt to each student's learning pace and style. Natural language processing allows for more nuanced interactions between students and digital learning platforms. Machine learning algorithms can analyze vast amounts of educational data to identify trends and optimize teaching strategies. And let's not forget about AI-powered tools that can assist with grading, lesson planning, and administrative tasks, freeing up educators to focus on what they do best—inspiring and guiding their students.

But AI in education isn't only about making existing processes more efficient. It's about reimagining what's possible in the realm of learning. It's about creating immersive, interactive experiences that engage students on a whole new level. It's about personalizing education to an unprecedented degree, ensuring that every learner can reach their full potential.

The Revolutionary Potential of AI in Education: A Teacher's Journey

Meet Maria, a dedicated high school science teacher with 15 years of experience. Like many educators, Maria has always strived to bring out the best in her students but often felt constrained by the limitations of traditional teaching methods. That was until her school decided to implement AI-enhanced learning tools.

At first, Maria was skeptical. She wondered how a machine could possibly understand the nuances of teaching. But as she began to explore the possibilities, her perspective started to shift.

One day, while preparing for a lesson on cellular biology, Maria discovered an AI-powered virtual reality program. As she put on the VR headset, she found herself shrunk down to the size of a cell, floating through a richly detailed 3D model of human tissue. She could interact with organelles, trigger cellular processes, and even simulate the effects of different medications.

Maria was amazed by the experience. She immediately began imagining how her students would react and started planning ways to incorporate this immersive experience into her lessons.

But the AI's capabilities went beyond just creating engaging content. As Maria's students began using the system, she noticed how it adapted to each student's pace and learning style. Emma, who had always struggled with abstract concepts, was now excelling thanks to the visual and interactive approach. Meanwhile, Jamal, who usually finished assignments quickly and became bored, was being challenged with increasingly complex scenarios.

The AI didn't just present information—it asked probing questions, encouraging students to think critically and apply their knowledge in new ways. It provided instant feedback, allowing students to learn from their mistakes in real time. Maria found herself freed from the tedious task of grading multiple-choice tests, giving her more time for one-on-one interactions with students who needed extra support.

As the semester progressed, Maria was amazed by the insights the AI provided. It could predict which students might struggle with upcoming topics, allowing her to provide targeted support before problems arose. It even suggested connections between her biology lessons and concepts from other subjects, enabling exciting cross-disciplinary projects.

But perhaps most surprisingly, Maria found that AI was breaking down barriers she never thought possible. When a new student who spoke little English joined the class, the AI-powered translation tools allowed him to participate fully from day one while simultaneously supporting his language learning.

Reflecting on her experience, Maria realized that AI hadn't replaced her role as a teacher—it had enhanced it. She was no longer just a conveyor of information but a guide, mentor, and facilitator of more profound learning experiences.

Maria's perspective on teaching had transformed. She now saw her role not as having all the answers but as asking the right questions and guiding students to discover answers for themselves. And with AI as a partner, the questions they could explore seemed limitless.

Maria's experience is just one example of how AI is revolutionizing education. From personalized learning paths to immersive experiences, from real-time feedback to predictive analytics, AI is opening up possibilities that were once the realm of science fiction. And this is just the beginning. As we continue to explore and innovate, the classroom of tomorrow holds boundless potential.

Thinking Outside the Box: The True Promise of AI in Education

While these applications of AI in education are exciting, the true revolutionary potential lies in how AI can help us think outside the traditional educational box altogether. This book isn't just about using AI to do what we've always done, but better. It's about reimagining education from the ground up.

Consider this: Our current educational model, with its standardized curricula and age-based progression, is a relic of the industrial age. It was designed to produce workers for factories and offices, emphasizing conformity and the ability to follow instructions. But in the age of AI, where routine cognitive tasks can be automated, we need to foster different skills: creativity, critical thinking, embodied intelligence, and the ability to adapt to rapid change.

AI gives us the tools to break free from the one-size-fits-all model of education. It allows us to create learning experiences that are not just personalized, but truly individualized. Imagine a curriculum that adapts not just to a student's academic abilities, but to their interests, learning style, cultural background, and even their desire for learning on any given day.

But it goes beyond that. AI can help us tap into the enactive nature of learning—the idea that true understanding emerges not just from abstract thought but from the dynamic interaction between the learner and their environment. This approach, grounded in the work of Varela et al., emphasizes that cognition is fundamentally tied to the learner's embodied actions and their coupling with the world around them.

In this view, learning isn't about passively receiving information but about actively engaging with and enacting one's world. AI can facilitate this by creating rich, interactive environments that respond to and challenge the learner's actions, fostering a continuous cycle of perception and action that leads to deeper understanding.

This is where the concept of "embodied AI education" comes in. It's about using AI not just as a tool for information delivery but as a means to create rich, multi-sensory learning experiences. It's about leveraging technology to bring abstract concepts into the physical world, where students can interact with them in tangible ways.

Imagine students learning about ecosystems by tending to an AI-powered virtual forest, where every decision they make has cascading effects they can see and experience. Or consider a history lesson where students can have conversations with AI-powered historical figures, gaining not just factual knowledge but enactive experiential insights into different historical perspectives.

This approach to education isn't just about enhancing the learning process, though that's undoubtedly a valuable outcome. It's about fostering a profoundly different way of understanding the world and our place in it. By embracing the principles of chaos theory, we recognize that small changes in educational approaches can have far-reaching, often unpredictable effects—much like the proverbial butterfly effect.

Consider how a single insight gained through an embodied AI learning experience might ripple out to affect not just that student's understanding of a subject but their entire worldview. This, in turn, could influence their future decisions, their interactions with others, and ultimately, the shape of the world they help to unfold.

In this light, education becomes not just about preparing students to pass tests or even to navigate the world as it is. Instead, it's about empowering them to become active participants in a complex, dynamic system—our global society. It's about equipping them with the tools to understand and engage with the intricate, interconnected nature of our world, where small actions can lead to large-scale changes over time.

By adopting this chaos theory-inspired perspective, we shift our educational goals. We move beyond linear thinking about cause and effect and instead cultivate an appreciation for the subtle, nonlinear ways in which knowledge and actions can transform our reality. This prepares students not just for the world they'll inherit, but for the myriad possible worlds they might help bring into being through their understanding and actions.

Your Invitation to Reimagine Education

This book is an invitation—an invitation to explore, to question, to imagine. Throughout these pages, we'll delve into real-world case studies of AI in education. We'll hear from educators and students on the front lines of this revolution. We'll examine cutting-edge research and peek into the classrooms of the future.

But this book is more than that. It is a call to action. Whether you're an educator, a policymaker, a technologist, a small business owner, or simply someone who cares about the future of learning, you have a role to play in shaping this AI-empowered educational landscape.

So, are you ready to think outside the box? Are you ready to challenge your assumptions about what education can be? Are you ready to explore the frontiers of embodied AI education? Let's get started!

Chapter 1:

Embodied AI in Education—A

Paradigm Shift

How does a classroom where learning isn't just about absorbing information but about experiencing it with every fiber of your being sound? Where understanding complex concepts isn't a struggle against abstract ideas but a journey of discovery through your senses and interactions with the world around you. Welcome to the world of embodied AI in education—a place where the boundaries between mind, body, and environment blur and where learning becomes a holistic, immersive experience.

The Dawn of a New Learning Era

Let's start with a confession: I was once a staunch traditionalist when it came to education. Give me a well-worn textbook, a chalkboard, and a room full of attentive students, and I thought I had all I needed to impart knowledge. Oh, how wrong I was.

My awakening came on a crisp autumn morning when I watched my young nephew, Alex, struggle with the concept of photosynthesis. No matter how many times I explained it, no matter how many diagrams I drew, the idea just wouldn't stick. Frustrated and about ready to throw in the towel, I suggested we take a break and go for a walk.

As we strolled through the park, crunching fallen leaves beneath our feet, Alex suddenly stopped in front of a massive oak tree. He placed his hand on the rough bark and looked up at the canopy of golden leaves above us. "Uncle," he said, his eyes wide with wonder, "is this how trees eat sunshine?"

In that moment, something clicked. Alex wasn't just seeing a tree; he was feeling it, experiencing it with his whole body. In doing so, he had grasped the essence of photosynthesis in a way that no textbook explanation could have achieved.

This is the power of embodied learning. And when we combine this approach with the adaptive, responsive capabilities of artificial intelligence, we unlock a whole new realm of educational possibilities.

Defining Embodied Enactive Learning

Before we dive deeper, let's unpack the concept of "embodied enactive learning." While the term might seem complex, its essence is profoundly intuitive to our everyday experiences.

At its core, embodied enactive learning recognizes that our bodies play a crucial role in how we think and learn. Consider how you learned to ride a bike. Did you start by reading a manual on the physics of two-wheeled locomotion? Of course not. You got on the bike, felt the balance (or lack thereof), experienced the rush of movement, and gradually, through physical trial and error, mastered the skill. This is embodied learning in action.

The "enactive" part of the term emphasizes that we don't just passively receive information from our environment. Instead, we actively create our understanding through our actions and interactions. It's a two-way street: our actions shape our perceptions, and our perceptions guide our actions.

This approach is grounded in the seminal work of Varela et al. (1991) in "The Embodied Mind," which argues that cognition is deeply intertwined with our physical experiences and our interactions with the world around us. It aligns closely with Dewey's (1938) emphasis on experiential learning, as outlined in his influential work "Experience and Education." Dewey argued that genuine education comes through experience and that the quality of the experience is crucial for learning.

At the heart of this concept is the body-mind-world connection, extensively explored by Gallagher (2005) in "How the Body Shapes the Mind." This perspective posits that the interplay between our bodies, minds, and environment fundamentally shapes our cognitive processes. It challenges traditional views that separate the mind from the body and the world, instead proposing a deeply interconnected system of cognition.

This interconnected view aligns with and builds upon systems theory and environmental ecology. Capra (1996), in "The Web of Life," provides a comprehensive exploration of systems thinking, applying it to understanding living systems, including human cognition. Abram (1996), in "The Spell of the Sensuous," offers a compelling account of how our sensory perceptions connect us to the more-than-human world, further emphasizing the importance of our embodied connection to our environment.

These ideas are rooted in the foundational work of von Bertalanffy (1968), whose "General Systems Theory" laid the groundwork for understanding the interconnectedness of complex systems, including the relationship between learners and their educational environments.

In essence, embodied enactive learning views education not as a process of absorbing information but as a dynamic interaction between learner and environment, where understanding emerges through physical experiences and active engagement with the world.

The Theoretical Foundations: From Varela to the Present

To truly appreciate the revolutionary potential of embodied AI in education, we need to take a quick journey through its theoretical roots.

Don't worry—I promise to keep it as painless as possible. Think of it as the origin story of our educational superheroes.

Our tale begins in the early 1990s with a brilliant Chilean biologist, philosopher, and neuroscientist named Francisco Varela. Along with his colleagues Evan Thompson and Eleanor Rosch, Varela proposed what they called the "enactive approach" to cognition.

Varela and his team were swimming against the tide of traditional cognitive science, which viewed the mind as a kind of computer, processing information in a detached, disembodied way. Instead, they argued that cognition arises through a dynamic dance between an organism and its environment, mediated by the organism's sensory and motor capabilities.

In simpler terms, they were saying, "Hey, we can't understand how people think and learn if we ignore the fact that we have bodies and exist in a physical world!"

This idea wasn't entirely new. Varela's work built on earlier theories, including the phenomenology of Maurice Merleau-Ponty (1945), who emphasized the pre-reflective, embodied nature of human experience. Merleau-Ponty was all about how we experience the world directly through our bodies before we even start to think about it consciously.

But Varela and his colleagues took these ideas further. They incorporated insights from dynamical systems theory—a branch of mathematics that deals with complex, changing systems—and ecological psychology, which studies how organisms interact with their environments.

The result was a rich, multifaceted view of cognition that emphasized its embodied, enactive, and embedded nature. In this view, learning isn't just about acquiring information; it's about developing new ways of interacting with the world.

From Theory to Practice: Alex's Discovery

Now, let's bring this theory to life with a story. Remember Alex, my nephew who had that "aha" moment with the oak tree? Let's dive a bit deeper into his journey of discovery:

Alex had always been a bright kid, but traditional classroom learning often left him frustrated. Abstract concepts seemed to slip through his fingers like sand. But everything changed when his school implemented a new, embodied approach to science education.

One sunny morning, Alex's class ventured out to a nearby pond for their lesson on ecosystems. Instead of just reading about food chains and energy transfer, the students were equipped with tablet computers loaded with an AI-powered augmented reality (AR) app.

As Alex pointed his tablet at the pond, the app overlaid digital information onto the real-world view. He could see animated representations of microscopic organisms, watch the flow of energy through the ecosystem, and even fast-forward time to observe long-term changes.

But the magic really happened when Alex crouched down by the water's edge. The AI noticed his movement and prompted, "What do you think would happen if you touched the water?"

Intrigued, Alex dipped his finger in the cool pond. Immediately, ripples spread across the surface, both in reality and in the AR display. On his screen, Alex could see how his small action affected the entire ecosystem—startling small fish, stirring up nutrients, creating new currents.

"Whoa," Alex breathed, his eyes wide with wonder. "Everything's connected!"

In that moment, Alex wasn't just learning about ecosystems—he was experiencing them. He was using his body to interact with the environment, and the AI was helping him see the consequences of his actions in ways that would be impossible with the naked eye.

This is the power of embodied AI in education. It's not about replacing real-world experiences with virtual ones. It's about enhancing our interactions with the world, making the invisible visible, and helping us understand complex systems through direct, physical engagement.

Alex's story illustrates how embodied enactive learning, powered by AI, can transform education. It bridges the gap between abstract concepts and concrete experiences, making learning more engaging, more intuitive, and, ultimately, more effective.

As we continue our exploration of embodied AI in education, we'll see how this approach can be applied across various subjects and learning contexts. We'll delve into the intersection of embodied cognition and AI, explore how the body, mind, and environment interact in educational settings, and examine case studies of immersive, AI-enhanced learning experiences.

But before we do that, let's take a closer look at how AI is changing the landscape of embodied learning, opening up possibilities that were once the stuff of science fiction.

The Intersection of Embodied Cognition and AI

Now that we've laid the groundwork for understanding embodied enactive learning, let's dive into the exciting realm where this approach intersects with artificial intelligence. It's at this crossroads that the real magic happens, transforming the way we think about education and learning.

The integration of embodied cognition principles with artificial intelligence opens up new possibilities for education. This intersection is explored in depth by Pfeifer and Bongard (2006) in their book "How the Body Shapes the Way We Think: A New View of Intelligence," which argues for an embodied approach to AI that takes into account the crucial role of physical embodiment in cognitive processes.

Imagine AI not as a replacement for human cognition but as an extension of our embodied experiences. This is where we're headed in education, and it's a game-changer.

AI as an Embodiment Enhancer

Traditional AI often focuses on mimicking human cognitive processes, like problem-solving or pattern recognition. But when we combine AI with embodied learning principles, we create systems that can enhance and extend our physical interactions with the world.

Let's consider a simple example: learning to play a musical instrument. Traditionally, you might learn by reading sheet music, watching an instructor, and practicing. Now, imagine an AI-enhanced violin that can sense your movements, analyze your sound, and provide real-time feedback not just through visual or auditory cues but through subtle vibrations or resistance in the bow. This AI isn't replacing the physical experience of playing; it's augmenting it, making the learning process more intuitive and embodied.

Adaptive Learning Environments

One of the most powerful applications of AI in embodied learning is the creation of adaptive environments. These are spaces—physical or virtual—that can change in response to a learner's actions and needs.

Picture a classroom where the physical space itself is responsive: Smart walls that can display information based on a student's gaze or gestures. Floors that can change texture or resistance to simulate different environments. AI-powered projections that can turn any surface into an interactive learning tool.

These aren't simply cool gadgets; they're tools that can profoundly enhance embodied learning experiences. They allow students to interact with complex concepts in physical, tangible ways, adapting to each learner's unique needs and learning style.

The Body-Mind-World Connection

At the heart of embodied enactive learning is the idea that cognition emerges from the interplay between body, mind, and environment. AI can help us visualize and interact with this interplay in unprecedented ways.

Consider a biology lesson on the human circulatory system. With AI-powered augmented reality, students could see a 3D representation of their own circulatory system overlaid on their bodies. As they move, run, or change posture, the AI could show in real time how their heart rate changes, how blood flow shifts, and how different organs are affected.

This isn't just a visual aid; it's a whole-body learning experience. Students aren't memorizing facts about the circulatory system; they're experiencing it, feeling it, and seeing how it responds to their actions. This kind of embodied understanding goes far beyond what traditional teaching methods can achieve.

Exploring the Body-Mind-Environment Interaction in Educational Contexts

Now that we've seen how AI can enhance embodied learning, let's explore more deeply how the body, mind, and environment interact in educational settings and how AI can facilitate these interactions.

The Classroom as an Ecosystem

First, let's reconceptualize the classroom—not as a static space with desks and a blackboard, but as a dynamic ecosystem where learning emerges from the interactions between students, teachers, physical space, and technology.

In this ecosystem, AI acts as a facilitator, helping to orchestrate these interactions for optimal learning. It might adjust lighting and temperature based on students' energy levels, suggest reconfigurations of the physical space for different activities, or create personalized ambient soundscapes to enhance focus or creativity.

But it goes beyond just environmental control. AI can help create what we might call "cognitive affordances"—opportunities for learning that arise from the interaction between learner and environment.

For example, imagine a history lesson on ancient civilizations. As students move around the classroom, AI-powered projections could transform the space into a virtual archaeological dig site. Students could use their bodies to "excavate" artifacts, with the AI providing contextual information based on their actions and discoveries. This isn't just making learning fun (although it certainly does that); it's creating a rich, multi-sensory learning environment that engages the whole body and mind.

Embodied Problem-Solving

One of the most powerful applications of embodied AI in education is in problem-solving and abstract thinking. Traditional education often treats problem-solving as a purely mental activity, but embodied cognition tells us that our physical experiences shape our cognitive processes.

AI can help bridge the gap between abstract problems and physical experiences. Let's take a math example: understanding geometric transformations like rotations and reflections. An AI system could create an augmented reality environment where students physically move shapes around, using their bodies to perform transformations. The AI could track their movements, provide feedback, and gradually introduce more complex concepts based on their embodied understanding.

This approach taps into our innate spatial reasoning abilities, grounding abstract mathematical concepts in physical experience. It's not about making math "easier"; it's about making it more intuitive, more connected to our embodied understanding of the world.

Social Learning and Embodied AI

Learning isn't just an individual process; it's deeply social. Our bodies play a crucial role in social interaction and communication, and AI can help enhance these embodied social learning experiences.

Imagine a language learning scenario where students are tasked with having conversations in a new language. An AI system could analyze not just their words but their body language, facial expressions, and tone of voice. It could provide real-time feedback on pronunciation, suggest culturally appropriate gestures, or even simulate different social contexts for practice.

This kind of embodied social learning goes far beyond traditional language drills. It recognizes that language is not just about vocabulary and grammar but about the whole-body experience of communication.

Immersive Experiences: Case Studies in Embodied AI Learning

To really understand the potential of embodied AI in education, let's look at some case studies of immersive learning experiences. These examples show how the principles we've discussed can be applied in real educational settings.

Case Study 1: The Virtual Biosphere

At a high school in Tokyo, students are using a cutting-edge AI-powered virtual reality system to study ecology. The system, called the Virtual Biosphere, allows students to immerse themselves in various ecosystems around the world.

But this isn't just a fancy 3D nature documentary. As students move through the virtual environment, they can interact with plants and animals, change environmental variables, and see the long-term effects of their actions play out in accelerated time.

The AI adapts the experience to each student's interests and learning pace. If a student shows a particular fascination with a certain species, the system might generate more in-depth information or create challenges related to that species' conservation.

What's truly revolutionary about the Virtual Biosphere is how it engages the whole body in learning. Students might need to crouch down to examine soil samples, reach out to touch virtual plants (with haptic feedback gloves providing realistic sensations), or even mimic animal movements to understand their behavior better.

Teachers report that students using the Virtual Biosphere not only show a better understanding of ecological concepts but also develop a deeper connection to the natural world. They're not just learning about ecosystems; they're experiencing them, fostering a sense of stewardship that extends beyond the classroom.

Case Study 2: The AI Chemistry Lab

At a university in California, chemistry students are using an AI-enhanced laboratory system that's revolutionizing how they learn about chemical reactions.

The system uses a combination of augmented reality and haptic feedback to create a safe, immersive environment for experimenting with chemical reactions. Students can "handle" virtual molecules, feeling their structures through haptic gloves. They can combine elements and compounds, with the AI simulating reactions in real-time.

What's particularly impressive about this system is how it bridges the gap between the macroscopic and microscopic worlds. Students can zoom in to see molecular interactions and then zoom out to observe large-scale effects. The AI guides this process, highlighting important details and explaining phenomena as they occur.

But it's not just about visualization. The system also incorporates embodied learning principles in clever ways. For example, when learning about reaction kinetics, students might physically act out the roles of molecules, with the AI tracking their movements and using them to model reaction rates.

Professors have noted that this embodied, immersive approach helps students develop an intuitive understanding of chemical principles that goes beyond mere memorization of formulas and reactions.

Case Study 3: Learning in Digital Worlds

A public charter high school in the Greater Boston area is using virtual reality (VR) technology to revolutionize how students learn and engage with science and engineering concepts. This innovative approach, studied by Eileen McGivney, a Ph.D. candidate in Human Development, Teaching, and Learning at Harvard Graduate School of Education, combines immersive experiences with educational objectives to enhance student learning and self-perception (Bauld, 2021).

Students use VR headsets to explore and interact with virtual environments relevant to their studies. In one project, students in a civil engineering course observed different structures around the world, visiting the pyramids in Egypt and soaring over skyscrapers in New York City. This immersive technology allows students to experience places and concepts that would be impossible or impractical to visit in person.

What sets this system apart is its focus on creating powerful, emotional experiences for students while delivering real educational benefits. The VR experiences are designed to instill an increased sense of competence and motivation in students. Many of the participants are English-language learners and first-generation Americans who have particularly appreciated the ability to visit and share locations that hold personal meaning to them (Bauld, 2021).

The system adapts to each student's needs and background. Teachers can observe how students engage with the material through these immersive experiences, gaining insights into their learning processes. McGivney's research indicates that students using VR technology report better ability to focus and a greater connection to the material.

Educators using this VR-based learning approach report that students develop a much deeper, more nuanced understanding of scientific and engineering concepts. They're not just memorizing facts; they're developing a rich, embodied understanding of complex ideas through direct experience. The technology has shown promise in helping students see themselves as scientists, potentially influencing their future career choices.

While the technology offers exciting possibilities, McGivney emphasizes that VR should not be seen as a replacement for traditional classroom learning. Instead, her research aims to understand how this technology can best complement and enhance existing educational systems, providing powerful tools for engagement, representation, and learning (Bauld, 2021). These case studies illustrate the transformative potential of embodied AI in education. By creating immersive, whole-body learning experiences, we can help students develop deeper, more intuitive understandings of complex subjects. We're not just changing how students learn; we're changing their relationship with knowledge itself.

As we move forward, we'll explore more applications of embodied AI across different subjects and age groups. We'll also delve into the challenges and considerations of implementing these technologies in educational settings. But for now, take a moment to imagine the possibilities. How might your own learning experiences have been different if you had access to these kinds of embodied AI tools? More importantly, how might we reshape education for future generations with these powerful new approaches that emphasize enaction and embodied cognition?

Challenges and Ethical Considerations

As exciting as the prospects of embodied AI in education are, it's crucial that we approach this new frontier with our eyes wide open. Like any transformative technology, embodied AI in education comes with its own set of challenges and ethical considerations. Let's dive into some of these issues:

The Digital Divide: Ensuring Equitable Access

One of the most pressing concerns is the potential to exacerbate existing educational inequalities. The kind of immersive, AI-powered learning experiences we've discussed require significant technological infrastructure. There's a real risk that only well-funded schools in affluent areas will have access to these tools, leaving other students behind.

This digital divide isn't just about hardware. It's also about the skills and support needed to effectively implement these technologies. Teachers need training, schools need technical support, and curricula need to be adapted. Addressing this challenge will require concerted effort from policymakers, educators, and technology providers to ensure that embodied AI enhances educational opportunities for all students, not just a privileged few.

The Role of the Teacher

As AI takes on more roles in the classroom, from providing personalized feedback to creating adaptive learning environments, we need to carefully consider the changing role of human teachers.

The goal of embodied AI in education should never be to replace teachers but to empower them. AI can take over routine tasks, freeing teachers to focus on what they do best: inspiring students, providing support, and guiding complex discussions.

However, this shift will require a reimagining of teacher training and professional development. Teachers will need to become adept at working alongside AI systems, interpreting AI-generated insights, and designing learning experiences that leverage the strengths of both human and artificial intelligence.

Maintaining Human Connection

One of the core tenets of embodied cognition is the importance of social interaction in learning. As we integrate more AI into education, we need to be careful not to lose the vital human connections that are so crucial to effective learning.

There's a risk that students might become overly reliant on AI feedback and guidance, potentially at the expense of peer-to-peer learning and student-teacher relationships. We need to design embodied AI systems that enhance, rather than replace, human social interactions in the learning process.

The Future of Embodied AI in Education

Despite these challenges, the potential of embodied AI to transform education is enormous. As we look to the future, several exciting prospects emerge:

Personalized Learning Pathways

As AI systems become more sophisticated at understanding individual students' learning styles, strengths, and weaknesses, we can envision truly personalized learning pathways. These would adapt content not just to students' academic levels but also to their physical states, interests, and personal goals.

Imagine a student who learns best through movement being guided through a physics lesson via an AI-powered dance routine, while another student who thrives on social interaction might learn the same concepts through an AI-facilitated group project.

Seamless Blending of Physical and Digital

The line between physical and digital learning environments will likely become increasingly blurred. We might see classrooms with smart surfaces that can instantly transform into interactive learning tools or AI-powered holographic displays that can bring abstract concepts to life in three dimensions.

Wearable technology could play a big role here, with smart glasses or haptic feedback suits allowing students to interact with digital information as if it were part of their physical environment.

Adaptive Enactive Learning Systems

As our understanding of enactive cognition deepens, we can expect to see AI systems that are more attuned to students' embodied interactions with their environment. These systems might adjust the pace or style of instruction based on a student's physical engagement and sensorimotor patterns or provide targeted support when a student's actions indicate difficulty with a concept.

For example, an AI system might detect that a student is struggling with a mathematical concept by observing their gestures and interactions with virtual objects. It could then adjust the learning environment to provide more tangible, interactive representations of the concept, allowing the student to engage with it in a more embodied way.

This approach could be particularly powerful in supporting students with special educational needs. By adapting learning experiences to individual sensory preferences and motor capabilities, these systems could create more inclusive educational environments that cater to diverse learning needs.

Imagine a physics lesson where students with different physical abilities can all participate fully, with the AI system adapting the interactive elements to each student's unique way of engaging with the world. One student might use large arm movements to interact with virtual objects, while another uses subtle finger gestures, but both are able to explore and understand the concepts through their own embodied experiences.

These adaptive enactive learning systems wouldn't just be reacting to students' actions; they would be actively creating opportunities for meaningful sensorimotor interactions that enhance understanding. They could dynamically generate learning scenarios that challenge students to use their bodies and their environment in new ways to grasp complex ideas.

As we move forward, developing such systems will require a deep understanding of embodied cognition and enactive learning principles. This will challenge us to rethink not just how we present information but also how we create environments and interactions that allow enactive learning to emerge through embodied experience.

Lifelong Learning Support

The principles of embodied AI learning are applicable to more than traditional classroom settings. As the pace of technological change accelerates, there's an increasing need for lifelong learning and skill development.

We might see AI systems that support embodied learning throughout our lives, from professional development courses that use virtual reality to simulate workplace scenarios to language learning apps that use augmented reality to turn everyday environments into immersive language practice opportunities.

Cross-Cultural and Global Learning

Embodied AI could also revolutionize how we approach global and cross-cultural education. Imagine virtual exchange programs where students can "visit" classrooms around the world, not just through video calls but through immersive, embodied experiences that allow them to interact with peers in virtual shared spaces.

Or consider how AI could facilitate language learning by creating embodied experiences that go beyond vocabulary and grammar to include cultural gestures, proxemics, and other non-verbal aspects of communication.

Embracing the Embodied AI Revolution in Education

As we stand on the brink of this educational revolution, it's natural to feel a mix of excitement and apprehension. The challenges we face in implementing embodied AI in education are significant, but so too are the potential benefits.

By embracing an approach to education that recognizes the deep connections between body, mind, and environment and leveraging the power of AI to enhance these connections, we have the opportunity to create learning experiences that are more engaging, more effective, and more inclusive than ever before.

The key will be to move forward thoughtfully, always keeping the needs and well-being of learners at the center of our innovations. We must strive to create AI systems that enhance human capabilities rather than replace them, that foster connection rather than isolation, and that expand opportunities for all learners, not just a privileged few.

As educators, policymakers, technologists, and learners, we all have a role to play in enacting this transformative future of education. Our task is not merely to implement new technologies but to fundamentally reimagine learning as an embodied, enactive process that unlocks human potential. We must strive to create educational environments that allow learners to be the best they can be, fostering a state of embodied flow as a way of being and knowing.

This approach challenges us to move beyond traditional notions of education. Instead of asking how we can use AI to improve existing educational models, we should ask: How can we create learning experiences that enable students to enact their understanding through meaningful interactions with their environment? How can we design educational spaces—both physical and virtual—that invite students into embodying flow, where learning becomes a natural, embodied process of engaging with the world?

Our role is to open up learning, to transform it from a passive reception of information into an active, embodied exploration of possibilities. We must continuously reimagine what's possible when we bring together human potential and artificial intelligence, not in service of predetermined learning outcomes but in service of expanding the horizons of human capability and understanding.

This vision requires us to ask profound questions about the nature of knowledge, the process of learning, and the purpose of education. It demands that we look beyond ethical and equitable implementation—though these remain crucial—to consider how we can create educational experiences that are inherently empowering, that allow each learner to enact their unique potential through embodied, flowing engagement with their world.

The embodied AI revolution in education is just beginning. As we move forward, let's do so with curiosity, creativity, and a commitment to creating a future where every learner has the opportunity to engage with knowledge in deep, meaningful, and embodied ways. The classroom of the future is not just a place to absorb information but a space to foster potential, to create, and to grow - in mind-body transformative experiences.

In the following chapters, we'll examine specific applications of embodied AI across different subject areas, explore best practices for implementation, and consider how these technologies might reshape our educational institutions and practices. But for now, I invite you to pause and reflect. How might embodied AI change your own approach to teaching or learning? What possibilities excite you? What concerns do you have?

Chapter 2:

AI and the Body—A Catalyst for

Embodied Experiences

Imagine a classroom where students aren't just learning about the human heart—they're feeling it beat in their own chests, amplified and visualized through AI-powered sensors. Picture a history lesson where learners don't just read about ancient civilizations but physically build and interact with AI-enhanced models of long-lost cities. This is the new frontier of education, where artificial intelligence doesn't replace human experience; rather, it amplifies and enriches our bodily engagement with the world around us.

In this chapter, we'll explore how AI is revolutionizing education by embracing and enhancing our physical experiences. We'll explore the symbiotic relationship between our bodies and AI, examining how this partnership can catalyze deeper, more intuitive learning experiences.

Embracing the Role of the Body in AI Learning Processes

The integration of AI in education isn't about replacing physical experiences with virtual ones. Instead, it's about enhancing our bodily interactions with the world to create richer, more meaningful learning experiences. This approach is deeply rooted in the concept of embodied cognition, which posits that our physical experiences fundamentally shape our understanding of the world (Gallagher, 2005).

AI can serve as a bridge between our physical experiences and abstract concepts, making the intangible tangible. For instance, consider the challenge of teaching complex mathematical concepts like calculus. Traditional methods often rely heavily on abstract symbols and equations, which can be difficult for many students to grasp. But what if we could use AI to translate these concepts into physical, interactive experiences?

Imagine an AI-powered "math playground" where students can physically manipulate 3D representations of mathematical functions. As they move their hands to change the shape of a curve, they see and feel how this affects the area under the curve, providing an intuitive, bodily understanding of integration. This is not just theoretical—researchers at MIT have developed similar systems that allow students to "touch" mathematical concepts (Okuno et al., 2019).

This approach aligns with Varela et al.'s (1991) enactive approach to cognition, which emphasizes that our understanding of the world emerges from our sensorimotor interactions with it. By integrating AI into these interactions, we can expand the range of experiences available to learners, allowing them to engage with concepts in ways that were previously impossible.

The Neuroscience of Embodied Learning

To fully appreciate the potential of AI-enhanced embodied learning, it's crucial to understand the neurological basis for this approach. Recent advances in neuroscience have provided compelling evidence for the embodied nature of cognition, supporting the theoretical framework proposed by Varela et al. (1991) and elaborated by Gallagher (2005).

Research using functional magnetic resonance imaging (fMRI) has shown that when we think about or observe an action, the same neural pathways activate as when we physically perform that action (Rizzolatti & Craighero, 2004). This discovery of "mirror neurons" has profound implications for learning, suggesting that our understanding of concepts is intimately tied to our physical experiences.

Studies have demonstrated that physical movement enhances cognitive function. For instance, Hillman et al. (2009) found that aerobic exercise improves children's executive function and academic performance. This research underscores the importance of integrating physical activity into learning processes, a principle that AI-enhanced embodied learning can leverage and amplify.

By using AI to create rich, interactive physical experiences, we can potentially stimulate these neural pathways more effectively, leading to more profound and lasting learning. For example, an AI system could guide a student through a series of movements that mirror the concept being taught, whether it's the motion of planets in the solar system or the flow of electrons in a circuit, engaging both their motor cortex and their higher cognitive functions.

Expanding Sensory Learning With AI

While much of our discussion has focused on movement and touch, AI has the potential to enhance learning through all our senses. This multi-sensory approach aligns with the concept of "embodied minds in animated bodies" proposed by Varela et al. (1991), which emphasizes the role of all sensory experiences in shaping our understanding of the world.

Visual Learning

AI augmented reality (AR) and virtual reality (VR) technologies are revolutionizing visual learning. For instance, medical students can now use VR systems to "walk through" a 3D model of the human body, gaining a spatial understanding of anatomy that's impossible with traditional 2D textbooks. AI algorithms can adapt these visual experiences in real time based on the student's focus and learning pace, creating a personalized visual journey through complex concepts.

Auditory Learning

In the realm of auditory learning, AI is opening up new possibilities for understanding complex patterns through sound. Consider the field of data sonification, where numerical data is converted into sound. AI algorithms can transform vast datasets into auditory landscapes that students can explore, potentially revealing patterns that might be missed in visual representations. This approach could be particularly powerful for teaching concepts in fields like climate science or economics, where understanding large-scale trends is crucial.

Olfactory and Gustatory Learning

Even our chemical senses—smell and taste—can be engaged through AI-enhanced learning experiences. While we can't yet digitally transmit smells or tastes, AI can help create more immersive, context-rich environments that engage these senses. For example, in a history lesson about ancient trade routes, AI could guide students through creating historically accurate spice blends, linking the sensory experience to geographical and cultural information.

Case Study: AI Integration in Music Education

In a two-part series published in Music Teacher magazine, tech expert Tim Hallas explores the multifaceted integration of artificial intelligence in music education, revealing how this technology is reshaping classroom practices and student learning experiences. Hallas demonstrates that AI's role extends far beyond mere music creation, encompassing language processing, production assistance, and even administrative tasks. In the realm of language processing, large language models like ChatGPT have become valuable tools for students, allowing them to quietly seek explanations for complex musical terms like "counterpoint" without the potential embarrassment of asking in class (Hallas, 2023). These AI models also serve as essay planning aids, generating outlines that students can critique against mark schemes, thereby honing their critical thinking skills. Furthermore, students are learning to craft prompts for AI to create revision materials, fostering independent learning habits (Hallas, 2023).

Moving into music creation and production, Hallas highlights AI-based software that analyzes musical content and suggests riffs, chords, harmonies, and rhythms, proving particularly beneficial for students who might have a melody but struggle with accompaniment or drum parts (Hallas, 2024). In the production sphere, tools like Focusrite FAST processors employ AI to analyze audio and recommend effect settings, while Logic Pro's AI mastering plugin allows students to compare AI results with their own efforts (Hallas, 2024). The classroom use of vocal separation tools like lalal.ai for isolating parts in tracks further exemplifies AI's practical applications in music education (Hallas, 2024).

Beyond creative applications, AI is streamlining administrative tasks, with teachers using it to draft parent emails and generate seating plans, significantly reducing time spent on such duties (Hallas, 2023). However, Hallas doesn't shy away from addressing the ethical considerations surrounding AI use in education. He raises important questions about authorship and copyright when it comes to AI-enhanced work, notes the often basic quality of AI-generated essays, and emphasizes the enduring importance of human creativity in the music-making process (Hallas, 2024).

Ultimately, Hallas portrays AI not as a future prospect but as a present reality in music education, stressing the importance of teaching students to use AI effectively and safely. He envisions AI as a tool to enhance creativity rather than replace human input, demonstrating its potential to not only assist in music creation and production but also to develop critical thinking skills and streamline educational processes (Hallas, 2024). Through Hallas's observations, this case study illuminates the transformative impact of AI across various facets of music education, from creative processes to administrative tasks, while also prompting important discussions about the role of technology in art and education.

Challenges and Ethical Considerations

While the potential of AI-enhanced embodied learning is enormous, it's crucial to address the challenges and ethical considerations that come with this approach.

Digital Divide and Accessibility

One of the primary concerns is the potential to exacerbate existing educational inequalities. The kind of AI-enhanced learning experiences we've discussed often require sophisticated and expensive technology. There's a risk that only well-funded schools will have access to these tools, widening the gap between privileged and underprivileged students.

Moreover, we must consider accessibility for students with different physical abilities. While AI has the potential to create more inclusive learning experiences—for instance, by translating visual information into tactile feedback for visually impaired students—we must ensure that embodied learning approaches don't inadvertently exclude students with limited mobility.

Privacy and Data Security

The use of AI in education, particularly when it involves tracking students' physical movements and physiological responses, raises significant privacy concerns. How do we ensure that this sensitive data is collected, stored, and used ethically? What are the long-term implications of having such detailed data about students' learning processes and physical responses?

The Role of Human Teachers

As AI takes on a more significant role in facilitating learning experiences, we must carefully consider the changing role of human teachers. While AI can provide personalized feedback and create interactive experiences, it can't replace the empathy, creativity, and complex social interactions that human teachers bring to the learning process. The challenge will be to find the right balance, using AI to enhance rather than replace human teaching.

Maintaining Connection to the Physical World

While AI can create rich, immersive learning experiences, we must be cautious about overreliance on digital interfaces. There's a risk that in our enthusiasm for AI-enhanced learning, we might inadvertently disconnect students from direct, unmediated experiences of the physical world. As David Abram (1996) argues in "The Spell of the Sensuous," our sensory connection to the natural world is fundamental to our cognition and well-being. We must ensure that AI-enhanced learning deepens rather than diminishes this connection.

Future Directions: The Convergence of AI, Neuroscience, and Education

As we look to the future, the intersection of AI, neuroscience, and education holds immense promise. Advances in brain-computer interfaces (BCIs) could allow for even more direct integration of AI into our learning processes. Imagine a learning environment that adapts not just to our physical movements but to our brain activity in real-time.

Research into neuroplasticity—the nervous system's ability to form new neural connections—could inform the development of AI systems that optimize the timing and nature of learning experiences to promote the most effective neural growth throughout the body. This approach recognizes that cognition and learning are not confined to the brain but involve the entire nervous system. It could lead to deeply embodied learning experiences tailored to each learner's unique neurological patterns as they interact with their environment. By considering the whole nervous system, from the brain to the peripheral nerves, these AI-enhanced learning experiences could more fully engage the learner's entire body in the process of understanding and skill acquisition.

Furthermore, as our understanding of the gut-brain axis and the role of the microbiome in cognition grows, we might see AI systems that take into account the complex interplay between our digestive system, our immune system, and our cognitive function. This could lead to holistic learning environments that consider nutrition, physical activity, and cognitive challenges as part of an integrated approach to education.

Embodying the Future of Learning

As we've explored in this chapter, the integration of AI into education offers unprecedented opportunities to create embodied learning experiences. By embracing the role of the body through enactive learning, we can use AI as a catalyst to bring abstract concepts to life, making them tangible, interactive, and personally meaningful.

From soft robotics to AI-enhanced music education, these approaches demonstrate how technology can deepen our physical engagement with the world rather than diminishing it. They show us that in the age of AI, the body's enactive experience remains central to our understanding and our learning.

As we move forward, the challenge for educators, technologists, and policymakers will be to harness these powerful tools in ways that enhance learning for all students, regardless of their background or abilities. For transforming potential, we must strive to create AI-enhanced learning environments that are not only effective but also ethical, inclusive, and deeply human.

In the next chapter, we'll explore how these embodied AI experiences can be integrated into traditional classroom settings, examining the practical challenges and opportunities of this pedagogical shift. We'll look at how teachers are adapting their methods, how school systems are evolving to accommodate these new approaches, and how students are responding to these innovative learning experiences.

Chapter 3:

The Pedagogical Classroom Shift—

AI as a Teaching Partner

In the heart of Silicon Valley, a revolution is unfolding in Mrs. Rivera's fourth-grade classroom. The walls are lined with interactive screens, and each desk is equipped with a tablet. However, the most remarkable presence is that of Aiden, an AI teaching assistant designed to facilitate an embodied enactive learning environment.

How Educators Integrate Embodied Experience in AI Education

Mrs. Rivera's classroom exemplifies the cutting edge of AI integration in education. Here, learning is not confined to passive listening or reading; it's a full-body experience enhanced by AI technology.

As students file into the classroom, Aiden greets each one with a personalized message. For Lucy, who has been struggling with fractions, Aiden displays a quick game to reinforce yesterday's lesson. For Mark, who excels in reading, it offers a challenging word puzzle. This personalized approach sets the stage for a day of tailored, embodied learning experiences.

The integration of embodied experiences in AI education is grounded in the theory of embodied cognition, which posits that our cognitive processes are shaped by our physical experiences and interactions with the environment (Wilson, 2002). This theory comes to life in Mrs.

Rivera's classroom through AI-enhanced activities that engage students' whole bodies in the learning process.

For instance, when learning about geometry, students might use motion-capture technology to create shapes with their bodies. The AI system could then analyze their movements, providing real-time feedback and introducing more complex concepts as students master the basics. This approach not only makes abstract mathematical concepts more tangible but also leverages the connection between physical movement and cognitive understanding.

Enactive AI in the Learning Environment: Designing Interactive Embodied Activities

Today's lesson in Mrs. Rivera's class is about the water cycle. Instead of a traditional lecture, Mrs. Rivera activates Aiden's simulation module. The room transforms into a virtual ecosystem. Students reach out to touch the clouds, and as they do, rain begins to fall within the simulation. They stomp their feet, and the virtual ground absorbs the water, illustrating infiltration.

This immersive, interactive environment exemplifies enactive AI in education. The learning space responds to students' actions, creating a dynamic interplay between the learners and their environment. This approach aligns with the enactive cognition theory proposed by Varela et al. (1991), emphasizing the importance of sensorimotor interactions in cognitive processes.

The design of such interactive embodied activities requires careful consideration of both the physical space and the AI capabilities. In Mrs. Rivera's classroom, the AI system is integrated with various sensors and display technologies to create a responsive environment. Pressure-sensitive floors detect students' movements, while gesture recognition cameras track their interactions with virtual objects.

Beyond the classroom, enactive AI is being explored in outdoor learning environments as well. For example, the "Smart Garden" project at the University of Colorado Boulder uses AI-enhanced augmented reality to turn school gardens into interactive learning spaces. As students tend to plants, they can use tablets or smart glasses to see real-time data about soil moisture, plant growth, and local ecosystem interactions (Shapiro et al., 2017).

Real-Life Story: Enhancing Special Education With AI Assistants

While Mrs. Rivera's class shows the potential of AI in mainstream education, similar technologies are making significant impacts in special education. For example, the ECHOES project, developed by researchers at the University of Edinburgh, uses AI to create an interactive learning environment for children with autism. The system adapts to each child's individual needs, helping them develop social communication skills through personalized, embodied interactions (Porayska-Pomsta et al., 2012).

Another notable example is the CopyMe project, developed at the University of Sussex. This AI-powered system uses facial recognition technology to help children with autism learn to recognize and reproduce facial expressions. The system provides real-time feedback, celebrating successful attempts and offering gentle guidance when needed (Mavrikis et al., 2019).

These AI assistants are not meant to replace human teachers or therapists but to augment their capabilities. They can provide consistent, tireless support and collect detailed data on each child's progress, allowing human educators to tailor their interventions more effectively.

Importantly, these systems are designed with teacher efficiency in mind. For instance, the ECHOES system allows teachers to review a day's worth of student interactions in as little as 15 minutes, providing a quick yet comprehensive overview of each child's progress. The system generates easy-to-read reports, highlighting key areas of improvement and potential concerns.

Setting up new interactions or personalizing greetings for each student is equally straightforward. Teachers can use voice commands to adjust settings or create new scenarios, often taking no more than a few minutes. This can be done while performing other tasks, maximizing the use of prep time.

The CopyMe project goes a step further in minimizing prep time. Its AI continuously learns from each interaction, automatically adjusting difficulty levels and introducing new expressions based on the child's progress. This means teachers spend less time planning individual lessons and more time on high-value interactions with students.

Both systems offer flexible input methods: Teachers can use voice commands, touchscreens, or traditional keyboard inputs to interact with the AI, depending on their preference and the classroom situation. This flexibility allows educators to seamlessly integrate the technology into their existing routines.

By streamlining administrative tasks and providing rapid, actionable insights, these AI assistants allow special education teachers to focus more on what they do best0151—providing personalized support and building meaningful connections with their students.

Intelligent Tutoring Systems: Real-World Examples of AI Tutors

Aiden, the AI assistant in Mrs. Rivera's class, is an advanced example of an intelligent tutoring system (ITS). As the water cycle lesson progresses, Aiden monitors students' understanding through their interactions with the virtual environment. It adjusts the difficulty in real time, ensuring that each student is engaged at just the right level of challenge. For those who grasp the concept quickly, Aiden introduces variables like pollution and urbanization to explore.

This adaptive capability is a hallmark of effective ITS. Another real-world example is Carnegie Learning's MATHia platform, which has been implemented in schools across the United States. MATHia uses AI to provide step-by-step guidance in math problem-solving, adjusting its approach based on each student's performance (Pane et al., 2014).

The effectiveness of ITS lies in its ability to provide immediate, personalized feedback and adapt the learning path in real time. For instance, if a student consistently struggles with a particular type of problem, the system might break it down into smaller steps, provide additional examples, or even switch to a different explanation method that might better suit the student's learning style.

Advanced ITS are beginning to incorporate natural language processing capabilities, allowing for more nuanced interactions. For example, the AutoTutor system developed at the University of Memphis can engage students in dialogues about complex topics, asking probing questions and providing explanations tailored to the student's level of understanding (Graesser et al., 2012).

Case Studies of AI Differentiation

The personalized learning experiences in Mrs. Rivera's classroom demonstrate AI's potential for differentiation. Each student receives tailored challenges and support, allowing them to progress at their own pace.

This approach is being implemented on a larger scale in some educational systems. For instance, the AltSchool platform, developed in San Francisco, uses AI to track students' progress across various subjects and skills. The system generates personalized "playlists" of learning activities, allowing teachers to tailor their instruction to individual student needs (Hernandez, 2018).

Another interesting case study comes from the Riiid Labs in South Korea. Their AI tutor, Santa, has been used by millions of students preparing for the Test of English for International Communication (TOEIC). Santa uses deep learning algorithms to analyze each student's learning patterns and predict their test performance, allowing for highly personalized test preparation strategies (Riiid Labs, 2020).

In higher education, the University of Michigan has implemented M-Write, an AI-powered writing assistant. M-Write uses natural language processing to provide feedback on student essays, not just on grammar and style but also on the strength of arguments and the use of evidence. This allows for more frequent and detailed feedback on writing assignments, even in large classes (Schultz et al., 2020).

Educators' Perspectives: Challenges and Triumphs

While the potential of AI in education is exciting, it's not without challenges. Educators often need significant professional development to effectively integrate AI tools into their teaching. There are also concerns about data privacy and the digital divide, as not all schools have equal access to advanced AI technologies.

A survey conducted by the Education Week Research Center (2024) found that while many teachers are enthusiastic about the potential of AI in education, they also express concerns about its implementation. Common worries include the potential for AI to replace human teachers, the reliability of AI-generated insights, and the risk of over-reliance on technology.

Despite these challenges, many educators report positive outcomes. Mrs. Rivera, for example, finds that Aiden allows her to spend less time on routine tasks and more time on meaningful interactions with her students. She can focus on higher-order teaching aspects, like fostering critical thinking and providing emotional support, while Aiden handles personalized instruction and basic feedback.

Some teachers have found creative ways to address the challenges of AI integration. For instance, professional development programs have been established to help teachers understand and implement AI tools effectively. These programs provide educators with the knowledge and skills needed to integrate AI into their teaching methods, easing the burden on teachers and enhancing the learning experience for students. This approach not only supports teachers but also provides valuable leadership opportunities for students who assist in the integration process (Edutopia, 2021)

Role of Perception-Action Experience in Learning

The water cycle lesson in Mrs. Rivera's class illustrates the importance of perception-action experiences in embodied learning. As students interact with the virtual environment, touching clouds and stomping to see water absorption, they're not just observing the water cycle—they're experiencing it through their actions and perceiving the results in real time.

This tight coupling of action and perception enhances understanding by grounding abstract concepts in physical experiences. It's an approach that's being explored in various educational contexts, such as the PhysicPlayground developed at Arizona State University, where students learn physics concepts by drawing and interacting with simulated machines (Roscoe et al., 2013).

Research in cognitive science supports the importance of perception-action experiences in learning. For instance, studies have shown that gesturing while explaining a concept can help students understand and remember it better (Goldin-Meadow, 2011). Embodied AI learning environments can leverage this principle by creating opportunities for meaningful gestures and actions that are tightly linked to the concepts being learned.

The role of perception-action experiences in enactive learning extends beyond individual cognition to social learning as well. In Mrs. Rivera's class, students can see and interact with each other's actions in the virtual environment, creating opportunities for collaborative learning and peer teaching.

Reflection and Growth

At the end of the day in Mrs. Rivera's class, Aiden prompts the students to reflect on what they've learned. They dictate their thoughts to the AI, which helps them organize their ideas into a coherent summary. Mrs. Rivera reviews these reflections, gaining insights into her students' learning that she might have missed otherwise.

This reflective practice, enhanced by AI, helps consolidate learning and provides valuable feedback for both students and teachers. It represents a new frontier in assessment, where AI can help capture and analyze the nuances of each student's learning journey.

Embodied AI's role in facilitating reflection goes beyond simple record-keeping. Advanced natural language processing algorithms can analyze students' reflections to identify misconceptions, track the development of higher-order thinking skills, and even detect signs of perceptual engagement or frustration with the material.

By aggregating and analyzing reflections over time, the AI can provide insights into learning trends, both for individual students and for the class as a whole. This can help teachers identify areas where additional support might be needed or where the curriculum might be adjusted to better meet students' needs.

The Future of AI in Education

As we look to the future, the integration of AI in education is likely to become even more sophisticated and ubiquitous. Emerging technologies like brain-computer interfaces and advanced haptic feedback systems could create even more immersive and responsive learning environments.

For instance, researchers at the University of California, Berkeley are exploring the use of functional near-infrared spectroscopy (fNIRS) to measure brain activity during learning tasks. This technology could potentially allow AI systems to adapt not just to students' actions but to their cognitive states in real time (Soltanlou et al., 2018). Imagine a learning environment that can detect when a student is becoming frustrated or losing focus and automatically adjust the difficulty or presentation of material to re-engage them.

Another promising area is the use of AI in developing "serious games" for education. These games leverage the engaging qualities of video games to create immersive learning experiences. A prime example is Foldit, a game developed at the University of Washington, which uses AI to create puzzles that help players learn about protein folding (Cooper et al., 2010).

Protein folding is a complex biochemical process in which a protein structure assumes its functional shape. Understanding this process is crucial for advancing our knowledge of diseases and drug design. Many diseases, including Alzheimer's and Parkinson's, are associated with misfolded proteins. By learning about protein folding through Foldit, players are not just gaining abstract knowledge but contributing to real scientific progress.

For example, in 2011, Foldit players helped solve the structure of an enzyme involved in the reproduction of HIV, a puzzle that had stumped scientists for years. This breakthrough opened new avenues for the design of antiretroviral drugs. More recently, Foldit players have been working on designing proteins that could bind to and neutralize the coronavirus responsible for COVID-19.

The game demonstrates how AI can make complex scientific concepts accessible and engaging to a broad audience. AI algorithms in Foldit generate puzzles, provide real-time feedback, and even learn from players' strategies to improve the game's ability to assist in real scientific research.

This approach to education—using AI to create interactive, game-like experiences that connect directly to cutting-edge research—could revolutionize how we teach and learn about science. Students could potentially contribute to real scientific discoveries while learning, blurring the lines between education and research.

As these technologies continue to evolve, we can anticipate an educational landscape where learning is increasingly personalized, interactive, and directly connected to real-world applications. The classroom of the future might not just prepare students for the world - it might be a place where students actively shape that world through their learning experiences.

Embracing Embodied Symbiosis

As the final bell rings in Mrs. Rivera's classroom, she takes a moment to reflect on the day's lessons. The water cycle simulation that Aiden, her AI teaching assistant, had run was a resounding success. She remembers Lucy's eyes lighting up as she made the connection between the virtual rain she had created and the real-world weather patterns they had been studying.

Mrs. Rivera pulls up Aiden's end-of-day report on her tablet. In just a few minutes, she's able to review each student's progress, noting with satisfaction that Jamal, who usually rushes through assignments, spent extra time exploring the advanced pollution scenarios Aiden had introduced.

As she prepares for tomorrow, Mrs. Rivera realizes how much her teaching has evolved since Aiden became part of her classroom. She's no longer bogged down with grading multiple-choice tests or struggling to provide individualized attention to 30 students simultaneously. Instead, she's free to focus on what she loves most: inspiring curiosity, guiding discussions, and connecting with her students on a personal level.

She thinks back to her initial skepticism about AI in the classroom. Now, she sees Aiden not as a replacement but as a powerful partner, enhancing her ability to create engaging, effective, and inclusive learning experiences.

Tomorrow's lesson plan excites her. With Aiden's help, she'll be taking the class on a virtual field trip to the Amazon rainforest, exploring biodiversity and climate change. She knows that Mark, who's been struggling with engagement, will be thrilled with the interactive elements Aiden has prepared based on his interest in wildlife photography.

As Mrs. Rivera packs up for the day, she feels a sense of anticipation for the future of education. She knows that the symbiosis between human teachers and AI is just beginning to unfold. The challenge ahead is to continue harnessing these capabilities in ways that enhance, rather than replace, the crucial human elements of teaching and learning.

The goal, she realizes, is not to create AI-driven classrooms but to develop embodied AI learning environments where technology and human experience work in concert to unlock each student's potential. In her classroom, at least, that future is already becoming a reality.

As we move forward to the next chapter, we'll explore the ethical considerations surrounding AI in education. We'll learn how educators like Mrs. Rivera, along with policymakers and technologists, can ensure that as we embrace the potential of AI, we do so in a way that respects human dignity, promotes equity, and enhances experiential, enactive learning environments worldwide. This global perspective will challenge us to think beyond conventional educational paradigms, examining how AI can support diverse ways of learning and knowing across cultures and contexts. Mrs. Rivera's journey will continue to guide us, providing a real-world perspective on the challenges and opportunities that lie ahead in this exciting new frontier of education—

one that holds the potential not just to improve learning outcomes but to transform how we understand and interact with the world around us.

Chapter 4:

Ethics in Motion—Guiding

Principles for Embodied AI

As Mrs. Rivera enters her classroom one morning, she pauses to watch Aiden, her AI teaching assistant, greet each student with a personalized message. She marvels at how seamlessly the AI has integrated into her classroom, enhancing her ability to provide individualized instruction. But a nagging thought crosses her mind: With all the data Aiden is collecting on her students, how can she ensure their privacy and well-being are protected?

This question is at the heart of the ethical considerations surrounding AI in education. As we venture further into this new frontier of embodied AI and enactive learning environments, we must grapple with complex ethical dilemmas that arise from the intersection of technology, education, and human development.

Strategies and Success Stories Impacting Ethical Decisions

Ethical decision-making in AI education isn't only about avoiding pitfalls; it's also about proactively creating systems and practices that uphold our values and enhance human potential. Several strategies have emerged as particularly effective:

- **Transparent AI:** Schools that have implemented transparent, explainable AI systems have seen higher levels of trust from students, parents, and teachers. For example, the Finnish government initiated a nationwide project called "AuroraAI," aimed at promoting transparent AI that supports citizens throughout various life stages. While not directly tied to schools, this type of open, explainable system serves as a model for educational institutions to adopt, fostering trust by explaining how AI makes decisions (Ministry of Finance, 2020).

- **Collaborative Ethics Boards:** Establishing ethics boards that include educators, technologists, parents, and students can help institutions navigate ethical dilemmas in AI use. An example in higher education is Stanford University's Institute for Human-Centered AI, which has developed a collaborative AI ethics framework that includes input from diverse stakeholders. This kind of inclusive approach can be scaled down to K-12 settings (Stanford HAI, 2021).

- **Continuous Education:** Regular ethics training for all stakeholders has been a key element in cultivating a culture of responsibility. In the United Kingdom, the Alan Turing Institute has partnered with schools to develop curricula that educate students on AI and ethics, embedding ethical awareness into the learning process from an early age (The Alan Turing Institute, 2020).

- **Ethical AI Design:** Involving ethicists early in the AI design process is crucial. For instance, MIT's AI Ethics and Governance Initiative emphasizes the need for interdisciplinary collaboration in the development of AI systems. Although this applies to broader AI development, the principles can be applied to educational AI to ensure ethical design practices from the outset (MIT Media Lab, 2018).

- **Community Engagement:** Schools that engage their communities in discussions about AI ethics often develop more robust frameworks. In the United States, the University of Massachusetts Amherst hosts annual public forums on AI ethics, fostering community dialogue. While focused on higher education, this example shows how schools can engage the wider community in understanding AI's ethical implications

(UMass Amherst, 2020). One notable success story in education comes from Oulu, Finland, where schools implemented an AI system to assist with learning difficulties, involving students in designing the system's privacy settings. This led to increased trust and engagement, and the project has been praised for its ethical approach to AI integration in education (University of Oulu, 2021).

Ethical Considerations in AI Education

As we integrate AI more deeply into our educational systems, several key ethical considerations come to the forefront:

- **Autonomy and Agency**: How do we ensure that AI enhances rather than diminishes student autonomy and agency? This question became particularly poignant for Mrs. Rivera when she noticed some students becoming overly reliant on Aiden's suggestions, potentially stunting their independent thinking skills.

- **Equity and Access**: How can we prevent AI from exacerbating existing educational inequalities? The potential for AI to widen the gap between tech-rich and tech-poor schools is a growing concern in many educational circles.

- **Transparency and Accountability**: How do we make AI systems in education transparent and accountable? The "black box" nature of some AI algorithms can make it difficult to understand and challenge their decisions.

- **Human-AI Interaction**: How do we foster healthy relationships between students and AI systems? There's a delicate balance between leveraging AI's capabilities and maintaining the crucial human elements of education.

- **Cultural Sensitivity**: How can AI systems respect and adapt to diverse cultural contexts? AI systems trained primarily on data from one cultural context may not be appropriate or effective in others.

- **Data Privacy and Security**: How do we protect the vast amounts of sensitive data collected by educational AI systems? The potential for data breaches or misuse is a significant concern.

- **Emotional Impact**: How do we consider the emotional and psychological effects of AI on students? For instance, how might constant AI-driven assessment affect a student's self-esteem or stress levels?

Mrs. Rivera grapples with these questions daily. She's acutely aware that every decision Aiden makes—from which students need extra help to how to pace the lesson—has ethical implications. She's found that openly discussing these ethical considerations with her students helps address potential issues and provides valuable learning opportunities about digital citizenship and ethical technology use.

Bias and Fairness, Protecting Data Privacy

One of the most pressing ethical concerns in AI education is the potential for bias. AI systems, after all, are only as unbiased as the data they're trained on and the humans who design them.

Mrs. Rivera noticed this firsthand when Aiden seemed to be recommending more advanced math problems to boys than girls despite equal performance. This led to a school-wide audit of the AI's recommendation algorithms and a renewed commitment to fairness in their AI systems.

To address this, the school implemented several measures:

- **Diverse Data**: They ensured that the AI was trained on a diverse dataset representative of their student body.

- **Regular Audits**: They instituted regular audits of the AI's decisions to check for bias.

- **Bias-Detection Algorithms**: They implemented additional algorithms specifically designed to detect and mitigate bias in the AI's decision-making processes.

- **Human Oversight**: They established a system where teachers regularly review and can override the AI's recommendations if they detect bias.

Data privacy is another critical concern. The wealth of data collected by AI systems like Aiden—everything from academic performance to behavioral patterns—could be immensely valuable but also potentially harmful if misused.

To protect student privacy, Mrs. Rivera's school has implemented several safeguards:

- **Data Minimization**: They only collect data that is directly relevant to educational purposes.

- **Anonymization**: Where possible, they anonymize data to protect individual student identities.

- **Strict Access Controls**: They limit who can access student data and for what purposes.

- **Transparency**: They provide clear information to students and parents about what data is being collected and how it's being used.

- **Right to Be Forgotten**: They've implemented processes for students to request the deletion of their data after they leave the school.

Real-Life Story: Balancing Privacy and Innovation in School Settings

A real-world example of AI-driven facial recognition for attendance can be found in Telangana, India. The state government adopted this system to track attendance across 26,000 schools, replacing traditional methods. This initiative aimed to streamline the attendance process, reduce administrative burdens, and improve accuracy. The system uses an AI-based facial recognition application that captures students' facial details and matches them against stored data to mark attendance automatically.

Although this system brought significant efficiency and reduced the time teachers spent on administrative tasks, it raised privacy concerns. Key issues included data storage, potential misuse of biometric information, and the security of the collected data. The system was designed to make every transaction traceable to address these concerns and ensure transparency. This means that all actions taken within the system can be audited, providing a clear record of how data is used and accessed.

However, community involvement, such as the formation of privacy councils or committees to oversee the implementation and management of the system, was not a key feature of this initiative. This lack of direct community engagement has been a point of criticism, as involving stakeholders like parents, teachers, and privacy experts could enhance trust and address concerns more comprehensively.

Despite these challenges, the Telangana government's approach to integrating AI in school attendance systems highlights the potential benefits and complexities of using advanced technology in educational settings. It underscores the importance of balancing innovation with robust privacy protections to gain the trust of the community and ensure the ethical use of technology (Privacy International, 2023; AI-Scholar, 2023).

Designing Curriculum to Include Students Working on Ethical Solutions

Many schools are incorporating AI ethics into their curriculum, recognizing that today's students will shape the ethical landscape of tomorrow's AI.

In Mrs. Rivera's class, students don't just use AI—they critically engage with it. She's designed projects where students analyze the ethical implications of AI decisions, propose improvements to AI systems, and even code their own ethical AI algorithms.

One particularly successful project involved students creating an "ethical AI manifesto" for their classroom. This not only deepened their understanding of AI ethics but also gave them a sense of ownership over the ethical framework governing their learning environment.

The manifesto included principles such as:

- **Transparency**: All AI decisions should be explainable and open to questioning.

- **Fairness**: AI systems should be regularly checked for bias and corrected when found.

- **Privacy**: Student data should be protected, and students should have control over their personal information.

- **Human-Centric**: AI should enhance, not replace, human interaction in education.

- **Continuous Improvement**: The ethical framework should be regularly reviewed and updated as technology evolves.

This project had a ripple effect, inspiring the school board to adopt a similar approach to developing district-wide AI policies.

Educator's Reflection: Ethical Dilemmas and Resolutions for Inclusive Education

As the school year draws to a close, Mrs. Rivera reflects on the ethical challenges she's navigated. There was the time Aiden identified a student as potentially having a learning disability—how to handle that sensitive information? Or the instance when a student figured out how to "game" Aiden's recommendation system—how to maintain the system's integrity without stifling the student's creativity?

Each dilemma required careful consideration, open communication, and, often, innovative solutions. Mrs. Rivera found that involving students in these ethical discussions not only led to better solutions but also provided invaluable learning experiences.

In the case of the potential learning disability, Mrs. Rivera decided to approach the student and their parents privately, presenting Aiden's observations as one piece of information to consider rather than a definitive diagnosis. This led to a productive conversation about learning styles and potential support strategies without stigmatizing the student or overstating the AI's role.

For the student who "gamed" the system, Mrs. Rivera turned it into a teachable moment about AI limitations and the importance of integrity. She worked with the student to document how they discovered the loophole, turning their mischief into a valuable contribution to improving the AI system.

These experiences have led Mrs. Rivera to develop a framework for addressing ethical dilemmas in an embodied AI-enhanced education:

1. **Identify the Stakeholders**: Consider who is affected by the AI's decisions and actions.

2. **Clarify the Ethical Principles at Stake**: Is it a question of privacy, fairness, transparency, or something else?

3. **Gather Information**: Understand the full context of the situation and the AI's role in it.

4. **Consider Alternatives**: Brainstorm different approaches to addressing the dilemma.

5. **Involve the Community**: When appropriate, engage students, parents, and colleagues in the decision-making process.

6. **Make a Decision and Explain It**: Choose the best course of action and be transparent about the reasoning.

7. **Monitor and Adjust**: Keep track of the outcomes and be willing to revise the approach if needed.

She's come to see ethics not as a set of rigid rules but as an ongoing, enactive dialogue—a continuous process of questioning, learning, and adapting. In her view, the goal isn't to create a perfect, ethically infallible AI system but to foster a learning environment where ethical considerations are at the forefront, mistakes are viewed as opportunities for growth, and the focus always remains on enhancing human potential.

The Road Ahead: Emerging Ethical Frontiers

As AI continues to evolve, new ethical frontiers are emerging. Mrs. Rivera and her colleagues are already beginning to grapple with questions such as:

- **Emotional AI**: As AI systems become better at recognizing and responding to human emotions, how do we ensure they're used ethically in educational settings?

- **AI-Generated Content**: When AI can generate essays, art, and even code, how do we redefine academic integrity and assessment?

- **Predictive Analytics**: If AI can predict a student's academic trajectory with high accuracy, how do we use that information ethically without creating self-fulfilling prophecies?

- **AI Rights**: As AI systems become more advanced, will we need to consider their rights and ethical treatment?

- **Global AI Governance**: How do we ensure ethical AI use in education across different cultural and regulatory contexts?

These questions don't have easy answers, but Mrs. Rivera recognizes that they are part of the ever-evolving landscape of embodied AI education. She sees her classroom not as a static environment but as a living, breathing ecosystem of learning, where ethical considerations are continuously enacted and re-enacted through daily interactions and decisions.

This approach aligns with the principles of flow pedagogy, where learning is seen as a dynamic process of constant adaptation and growth. In Mrs. Rivera's classroom, ethical awareness isn't a fixed set of rules but a fluid, responsive framework that evolves in tandem with technological advancements and emerging challenges. It's an embodied practice where students don't just learn about ethics but experience and enact ethical decision-making in real time.

As we continue to develop and implement AI in education, Mrs. Rivera's experiences remind us that ethics in AI education isn't a destination but a journey. It's about creating an enactive learning environment that opens up new possibilities, constantly challenging us to reconsider and refine our ethical stance. This approach embodies the essence of AI education—a continuously changing, adaptive process that responds to and shapes the world around it.

In the next chapter, we'll explore how this dynamic, embodied approach to ethics plays out in interdisciplinary education. We'll examine how AI can bridge gaps between different subjects, creating a holistic learning experience that respects the unique ethical considerations of each field while fostering a unified ethical consciousness. From STEM to humanities and beyond, we'll see how the ethical foundation laid in embodied AI classrooms like Mrs.

Rivera's can be applied and adapted, creating a rich tapestry of ethical awareness that spans the entire educational spectrum.

This journey into interdisciplinary AI education will further illuminate the potential of embodied learning, showing us how an ethically grounded, flow-based approach can open up unprecedented possibilities for education in the AI age.

Chapter 5:

Bridging the Gap—AI In

Interdisciplinary Education

In a bustling classroom at the crossroads of technology and creativity, Mr. Chen's students are embarking on a journey that defies the traditional boundaries of education. Here, art meets algorithm, and history harmonizes with code. This is the domain of interdisciplinary education, supercharged by the power of embodied AI.

Retention in Learning: Promoting Creativity for Innovation

As Mr. Chen observes his students immersed in their projects, he reflects on the transformative power of interdisciplinary AI education. By breaking down the silos between subjects, students are not simply memorizing facts—they're forging connections, thinking critically, and solving problems in innovative ways.

One of Mr. Chen's students, Alex, struggled with traditional math classes. However, Alex's interest was piqued when introduced to an AI-powered design tool that uses mathematical principles to create 3D sculptures. The project's tactile, visual, embodied nature helped Alex grasp complex geometric concepts that had previously seemed abstract and irrelevant.

This approach aligns with recent research on retention in learning. A study published in the *Journal of Interactive Media in Education* found that students engaged in interdisciplinary, AI-enhanced projects showed a significant increase in long-term retention of key concepts compared to traditional single-subject approaches (Whitelock et al., 2018).

Mr. Chen has structured his curriculum around what he calls the "CREATE" framework:

- **Connect**: Link concepts across disciplines.

- **Reflect**: Encourage embodied metacognition about the learning process.

- **Experiment**: Use AI tools to test hypotheses and iterate designs.

- **Analyze**: Apply data science techniques to various subjects.

- **Transform**: Use insights to create novel solutions.

- **Evaluate**: Critically assess the implications of AI-generated possible outcomes.

This framework has not only boosted retention but has also fostered a spirit of innovation among his students. They're not just consumers of technology but active creators and problem-solvers.

The CREATE framework in action is evident in a recent project where students were tasked with designing a sustainable city. The project required them to integrate knowledge from urban planning, environmental science, economics, and social studies. Students used AI simulations to model the impact of their design decisions, from energy usage to traffic flow to social equity.

One group of students proposed a novel solution to urban heat islands by designing AI-controlled reflective surfaces that could dynamically adjust to weather conditions. This idea emerged from their analysis of data patterns in climate science, materials engineering, and urban design—a connection that might not have been made in a traditional, siloed approach to education. This approach embodies AI education in a transformative space, opening up new ways of knowing and interacting with the world. It utilizes multiple modalities across diverse subjects, connecting back to the theoretical concepts we introduced at the beginning. By integrating AI into education in this way, we create opportunities for students to engage with technology not just as users but as creators and critical thinkers. This aligns with the constructivist and connectivist learning theories we discussed earlier, emphasizing the importance of active learning, knowledge construction, and networked understanding in the digital age.

A Symphony of Data: Combining Arts and Sciences

The chapter opens with the story of Maya, a high school student with a passion for music and a curiosity for data science. With the help of an AI program named HarmonAI, she discovers patterns in classical compositions and translates them into visual data. This enactive approach to learning allows Maya to experience music theory through a new, multi-sensory dimension.

Maya's project is a prime example of how AI can bridge the gap between arts and sciences. As she feeds musical scores into HarmonAI, the program analyzes the compositions' structure, rhythm, and harmony. It then generates visualizations that represent these musical elements as colorful, interactive graphs and patterns.

For Maya, this process is revelatory. She's always loved music but never thought of it as mathematical before. Now, she can literally see the math in Mozart.

The project goes beyond mere analysis. Maya uses the insights gained from HarmonAI to compose her own pieces, creating a feedback loop between artistic expression and thus embodying data-driven composition. This blend of creativity and analytical thinking exemplifies the potential of AI in interdisciplinary education.

Maya's work has inspired other students to explore similar intersections between arts and sciences. A group of students is now using AI to analyze the linguistic patterns in Shakespeare's plays and sonnets, creating a "language fingerprint" for the Bard. They're using this to explore questions of authorship in disputed works, bringing together literary analysis, linguistics, and data science in a novel way.

Another student, inspired by Maya's work, is using AI to analyze the brush strokes in famous paintings. AI can identify patterns that are invisible to the human eye, revealing insights into the artists' techniques and even helping to authenticate disputed works. This project brings together art history, computer vision, and machine learning in an innovative way.

The Art of Algorithms: AI-Generated Historical Art

Next, we explore a project where students use AI to create digital art. They feed the AI descriptions of historical events, and it generates vivid images that capture the essence of those moments. As the students refine their inputs, the AI adapts, producing artwork that is both educational and emotionally resonant.

This project, led by Mr. Chen's colleague Ms. Patel, combines elements of history, language arts, and computer science. Students must research historical events, craft precise descriptions, and then interpret and refine the AI's output.

One student, Jamal, chose to focus on the moon landing. His initial description led to an AI-generated image that was technically accurate but was lacking....something. Through several iterations, Jamal refined his language, learning to balance factual details with evocative descriptions. The final image—showing an astronaut's boot print on the lunar surface with Earth reflected in the visor—was so powerful that it was featured in the school's art exhibition.

Ms. Patel notes that this project has dramatically increased engagement with historical topics. Students are no longer passive recipients of historical facts. They're actively interpreting and visualizing history, making the past come alive for them in a way she's never seen before.

The success of this project has led to its expansion into other subjects. In biology classes, students are using similar AI tools to visualize complex molecular processes. By describing protein folding or DNA replication in words and then seeing these processes rendered visually by the AI, students are gaining a deeper, more intuitive understanding of these microscopic phenomena.

In geography classes, students are using AI to generate images of future landscapes based on climate change projections. This not only helps them understand the potential impacts of global warming but also encourages them to think critically about the assumptions built into these projections.

These projects vividly illustrate the mind-body-world connection central to embodied AI education. By integrating AI into creative and analytical tasks, students are not just learning about subjects; they're actively engaging with and shaping their understanding through multi-sensory, interactive experiences. This approach exemplifies the transformative potential of embodied AI education, creating enactive learning environments where knowledge emerges through the dynamic interplay of physical interaction, cognitive processes, and technological augmentation.

From Maya's musical data visualizations to Jamal's AI-generated historical art, these projects demonstrate how embodied AI education can break down traditional subject boundaries, fostering a more holistic, interconnected approach to learning. They show how technology can be leveraged not just as a tool for information delivery but as a medium for creative expression and deep, experiential understanding.

These examples reinforce the theoretical foundations we discussed earlier, showing how embodied AI education can create rich, immersive learning experiences that engage students' minds and bodies in the process of knowledge construction. By doing so, they open up new possibilities for learning, encouraging students to explore, create, and innovate in ways that traditional educational approaches might not facilitate.

History in Holograms: Augmented Reality and Historical Figures

Mr. Chen's class then delves into the world of augmented reality, where AI brings historical figures to life as holograms. Students interact with these figures, asking questions and receiving answers based on historical texts. This immersive, embodying learning experience deepens their understanding of history and its relevance to the present.

The system, developed in collaboration with historians and AI experts, draws upon a vast database of historical documents, speeches, and biographical information. It uses natural language processing to interpret students' questions and generate responses that are not only accurate but also reflect the speaking style and personality of the historical figure.

During one session, students engage in a 'conversation' with a hologram of Marie Curie. They ask her about her discoveries, the challenges she faced as a woman in science, and her views on modern nuclear physics. Drawing upon Curie's writings and contemporaneous accounts, the AI provides nuanced responses that spark further questions and discussions.

This technology doesn't just make history more engaging; it helps students develop critical thinking skills. They learn to craft questions that elicit informative responses, to cross-reference the AI's answers with other sources, and to consider historical figures as complex, multifaceted individuals.

The success of the history holograms has inspired similar projects in other subjects. In literature classes, students can now interact with holographic representations of famous authors, discussing their works and creative processes. In science classes, holograms of pioneering scientists help students understand the historical context of major discoveries and the scientific method in action.

One particularly innovative application is in language classes, where students can practice conversations with AI holograms of native speakers from different historical periods. This helps with language skills and provides insights into how languages evolve over time.

Case Study: AI in Environmental Education at Cornell University

Cornell University has embarked on a groundbreaking initiative to tackle pressing environmental challenges through the integration of artificial intelligence (AI) in sustainability research. This initiative, supported by a substantial grant from the National Science Foundation's Research Traineeship Program, aims to train doctoral students in applying AI to critical areas such as sustainable materials, decarbonization of energy systems, climate-smart digital agriculture, and the global energy-food-climate nexus (Forward Pathway, 2023).

The program involves interdisciplinary collaboration among students from various fields:

- Biology students contribute data about local flora and fauna, which the AI uses to model population dynamics and biodiversity.

- Chemistry students analyze soil and water samples, feeding this data into the AI to track pollution levels and nutrient cycles.

- Physics students study energy flows within ecosystems, using AI-powered analytics to understand complex thermodynamic processes.

- Computer science students work on improving AI models and creating user interfaces for data visualization.

- Art students create visual representations of the data, turning complex ecological concepts into accessible infographics and animations.

- Social science students craft narratives based on simulated future scenarios, exploring the human impact of environmental changes.

The AI system doesn't just process this data contemporaneously; it creates a dynamic, evolving model of the ecosystem. Students can run simulations to see how changes in one area affect the entire system, fast-forward to see the long-term consequences of current actions, or rewind to understand how past events shaped the present environment.

This holistic approach has led to remarkable learning outcomes. Students have identified previously unrecognized threats to local wildlife, proposed innovative solutions for sustainable urban development, and even presented their findings to local government bodies. The program has had impacts beyond the classroom, with local conservation groups using the students' data and simulations to inform their work and city planning departments incorporating some of the students' sustainable development ideas into their long-term plans.

The success of this program has inspired similar initiatives in other institutions, focusing on different local issues. For example, one school in a coastal area is using a similar AI-driven approach to study marine ecosystems and the impact of climate change on sea levels, while another in an urban area is focusing on air quality and the urban heat island effect (Teachflow.AI, 2023).

Educator's Account: Fostering Cross-Disciplinary Thinking

Dr. Lydia Hernandez, the architect of the EcoSphere program, shares her insights on fostering cross-disciplinary thinking:

The key, she explains, is to create an environment where students don't just bring their subject-specific knowledge to the table but also learn to appreciate and integrate insights from other disciplines. AI acts as a bridge, providing a common platform where different types of data and ideas can interact.

Dr. Hernandez emphasizes the importance of collaborative projects that require input from multiple disciplines. When a biology student has to explain their findings to a computer science student to improve the AI model, both gain a deeper understanding of their own field and appreciate the value of other perspectives.

She also stresses the need for flexibility and creativity in assessment. Traditional exams often fall short in evaluating interdisciplinary skills. Instead, Dr. Hernandez advocates for project-based assessments that reflect real-world problem-solving scenarios, an embodied enactive approach to learning and teaching or pedagogy.

She says we're not just preparing students for existing jobs. We're equipping them with the skills to tackle complex, multifaceted challenges that we can't even imagine yet. The ability to synthesize knowledge from multiple disciplines, leverage AI tools effectively, and think critically about the implications of their work are the skills that will define success in the 21st century.

Dr. Hernandez has also developed a framework for promoting cross-disciplinary thinking, which she calls the "BRIDGE" method:

- **Boundary-Crossing**: Encourage students to explore beyond their comfort zones

- **Reflection**: Regular sessions for students to consider how different disciplines interact

- **Integration**: Projects that require synthesizing knowledge from multiple fields

- **Dialogue**: Structured discussions between students from different disciplines

- **Generation**: Creating new ideas or solutions that draw on multiple areas of expertise

- **Evaluation**: Assessing the strengths and limitations of interdisciplinary approaches or co-teaching with this approach and assessment

This framework has been adopted by several other schools and is now being studied by educational researchers as a model for fostering interdisciplinary thinking in the age of AI.

The Ethics of Integration

The chapter concludes with a reflection on the ethical considerations of integrating AI into interdisciplinary education. It discusses the importance of critical thinking, the challenges of ensuring equitable access, and the responsibility of educators to guide students through the complex landscape of embodied AI learning.

Mr. Chen and his colleagues are acutely aware of these challenges. They've implemented several strategies to address them:

- **Ethics Workshops**: Regular sessions where students discuss the implications of AI in various fields, from art creation to scientific research.

- **Transparency in AI**: Ensuring students understand how AI systems work, including their limitations and potential biases.

- **Equitable Access Programs**: Partnerships with local tech companies to provide necessary hardware and software to all students, regardless of economic background.

- **AI Literacy Curriculum**: Integrating lessons on AI fundamentals across all subjects, ensuring all students have a basic understanding of the technology they're using.

- **Human-AI Collaboration Emphasis**: Stressing that AI is a tool to enhance human creativity and problem-solving, not a replacement for human thought and effort.

The school has also established an AI Ethics Board composed of teachers, students, parents, and local technology experts. This board reviews all major AI implementations in the school, considering issues of privacy, fairness, and potential unintended consequences.

One of the board's key initiatives has been the development of an "AI Impact Assessment" framework. Before any new AI tool is introduced in the classroom, teachers must complete this assessment, considering questions such as:

- How might this AI tool advantage or disadvantage different groups of students?

- What data is the AI using, and how is student privacy protected?

- How transparent is the AI's decision-making process?

- What skills might students lose by relying on this AI tool?

- How can we ensure that students maintain agency and critical thinking when using this tool?

This framework has improved the ethical implementation of AI in the school and has become a valuable teaching tool. Students are involved in the assessment process, giving them practical experience in ethical decision-making and technology assessment.

As the bell rings and students file out of Mr. Chen's classroom, there's a palpable sense of excitement. These young learners are not just studying individual subjects—they're exploring the interconnected nature of knowledge itself, guided by AI but driven by their own curiosity and creativity. The classroom hums with the energy of newfound connections, each student carrying with them a piece of a larger puzzle that only becomes clear when viewed as a whole. Mr. Chen's innovative approach has transformed the learning space into a nexus of ideas, where the boundaries between disciplines blur and new understandings emerge from unexpected intersections. As they leave, the students carry with them not just facts and figures but a new way of seeing the world—one where every bit of knowledge has the potential to illuminate another, creating a rich tapestry of understanding that extends far beyond the classroom walls.

In the next chapter, we'll explore how the interdisciplinary skills developed in classrooms like Mr. Chen's are being applied in real-world settings, preparing students for the complex, AI-infused workplaces of the future. We'll examine case studies of recent graduates applying their interdisciplinary AI skills in fields ranging from environmental conservation to healthcare to creative industries. We'll also look at how universities and employers are adapting to this new breed of interdisciplinary, AI-savvy students, and what this means for the future of education and work.

Chapter 6:

The Enactive Playground—AI in

Extracurricular Activities

The halls buzz with excitement as the final bell rings at Westfield High School. For many students, the end of formal classes marks the beginning of a different kind of learning—one that takes place on sports fields, in music rooms, and in art studios. Here, in the realm of extracurricular activities, AI is opening up innovative frontiers of creativity, performance, and personal growth.

Embodying AI in Sports, Music, and Art Education

In the school gymnasium, the basketball team is wrapping up practice with the help of an embodied AI analysis system. Coach Thompson reviews the data on his tablet, which has been tracking each player's movements, shot accuracy, and overall performance throughout the session.

"Great improvement on your three-point shots, Jamal," Coach Thompson calls out. "The AI suggests adjusting your elbow angle slightly. Let's work on that tomorrow."

Meanwhile, in the music room, the school orchestra is rehearsing with an unusual addition to their ensemble—an AI conductor. The system, developed by a local tech startup, analyzes the musicians' playing in real time, offering suggestions on tempo, dynamics, and even emotional expression.

Down the hall in the art studio, students are exploring new realms of creativity with AI-assisted tools. Sarah, a senior known for her abstract paintings, is using a neural network to generate unique color palettes based on her own previous works. The AI doesn't replace her creativity but serves as a source of inspiration, challenging her to explore new combinations she might not have considered.

These scenes illustrate how AI is enhancing extracurricular activities across various domains, with implications that extend far beyond the school years into lifelong learning and future careers:

- In sports, AI-powered analytics are providing coaches and athletes with unprecedented insights into performance, helping to refine techniques and strategies. For instance:

  - In baseball, AI systems analyze batting stances and pitching motions, offering personalized recommendations to improve a player's swing or pitch accuracy. This technology, first developed for school teams, is now being adopted by professional leagues and could revolutionize player training and scouting.

  - In golf, AI-powered swing analysis tools provide instant feedback on form, club speed, and ball trajectory. Students learning golf today might find these skills valuable in future business settings, where golf often serves as a networking tool.

- In music, AI is serving not just as a tool for composition and arrangement, but as an interactive partner in the learning and performance process. This collaborative approach to creativity could translate into future innovations in fields like advertising jingle creation or soundtrack composition for video games and films.

- In visual arts, AI is opening up new avenues for creative expression, pushing the boundaries of what's possible and inspiring students to explore new techniques and styles. The skills developed in AI-assisted art creation could be applied in future careers in graphic design, user interface development, or even AI-human collaborative art installations.

These AI-enhanced extracurricular experiences are preparing students not just for potential careers in sports, arts, or music but for a future where human-AI collaboration is the norm across many industries. The ability to work alongside AI, interpreting and applying its insights creatively, will be a valuable skill in numerous professional contexts.

Embodied AI in Action: A School Band's AI Composer

The Westfield High Jazz Band has always been known for its innovative performances, but they're pushing the envelope even further this year. Under the guidance of their music teacher, Ms. Yamamoto, the band is collaborating with an AI composer named Genius Jam Tracks.

The project began when Ms. Yamamoto introduced the AI to her students as a creative challenge. "I want you to think of Genius Jam Tracks not as a replacement for your creativity," she told them, "but as a new instrument in our ensemble—one that can help us explore new musical territories."

The students fed Genius Jam Tracks with recordings of their previous performances, along with a selection of classic jazz pieces. The AI analyzed these inputs, learning the band's unique style and the fundamental structures of jazz music.

During one memorable session, lead trumpeter Chris was struggling with a particularly complex solo. Genius Jam Tracks suggested a series of note progressions that Alex had never considered. As he played through them, his eyes lit up with excitement.

"It's like the AI understands where I want to go with the music, even when I'm not sure myself," Chris exclaimed.

The collaboration culminated in a piece titled "Digital Rhapsody in Blue," a fusion of human and AI-generated music that became the centerpiece of the band's spring concert. The performance was met with a standing ovation, with audience members marveling at the seamless blend of traditional jazz elements and innovative, AI-inspired compositions.

This experience not only enhanced the students' musical skills but also sparked deep discussions about the nature of creativity and the role of AI in artistic expression. It embodied the principles of enactive learning, with students actively engaging with and responding to the AI's input, creating a dynamic feedback loop of musical exploration and growth.

Creative Expression through Generative AI Possibilities

The success of the jazz band's AI collaboration inspired other departments to explore the creative possibilities of generative AI. In the creative writing club, students began experimenting with AI-assisted storytelling.

Emily, a sophomore with a passion for science fiction, used a language model to co-write a short story. She would write a paragraph and then allow the AI to generate several possible continuations. Emily would then choose the one she found most interesting, edit it to fit her vision, and continue the process.

"It's like having a brainstorming partner that never gets tired," Emily shared. "Sometimes, the AI comes up with ideas that are completely out of left field, and that pushes me to think in new ways."

In the drama club, students used AI to generate character backstories and potential plot twists for their original play. This not only added depth to their performance but also taught them about narrative structure and character development.

The school's digital art class took things a step further by creating an "AI Art Gallery." Students used various AI tools to generate base images, which they then refined and transformed using traditional digital art techniques. The resulting exhibition was a stunning showcase of human-AI collaborative creativity.

These projects demonstrated how AI could serve not as a replacement for human creativity but as a tool to augment and inspire it. Students were learning not just to use AI but to engage with it critically and creatively—skills that would serve them well in an increasingly AI-integrated world.

Educator's Tale: The Joy of Learning With AI and the Unexpected

As the school year drew to a close, the teachers gathered to reflect on their experiences with AI in extracurricular activities. Mr. Reeves, the robotics club advisor, shared a particularly poignant story.

"When we introduced AI into our robotics projects, I expected it to help with things like optimal design and efficiency calculations," Mr. Reeves began. "What I didn't expect was how it would transform the way our students collaborate and problem-solve."

He went on to describe a project where students were tasked with building a robot that could navigate a complex obstacle course. The team used an AI simulation to test various designs before building them physically.

"What amazed me was how the AI's suggestions often led to heated debates among the students," Mr. Reeves continued. "They would argue passionately about why the AI's solution might or might not work in the real world. It was fostering critical thinking in a way I'd never seen before."

The unexpected experience came during the regional robotics competition. The team's robot encountered an unforeseen obstacle that wasn't in their original simulations. Instead of panicking, the students huddled together, discussing how their experience with the AI simulations had taught them to think adaptively.

"They treated the situation like a new simulation run," Mr. Reeves said, his voice filled with pride. "They quickly brainstormed potential solutions, weighing pros and cons just like they did with the AI's suggestions. In the end, they came up with a brilliant workaround that won them the competition."

This story highlighted how AI in extracurricular activities was doing more than just enhancing performance; it was opening unexpected possibilities for students in how to think, adapt, and collaborate in new ways.

As the teachers shared similar stories from their respective domains— sports, arts, music, and more—a common theme emerged. AI was not replacing the core aspects of these activities; rather, it was enhancing them and opening up new transformative possibilities for enactive learning and new pedagogical discoveries.

The discussion turned to the future, with teachers excitedly planning new ways to integrate embodied AI into their extracurricular programs. Ideas ranged from AI-assisted choreography for the dance team to AI-generated scenarios for the debate club.

As the meeting concluded, there was a palpable sense of anticipation for the coming school year. The teachers realized that they were not just teaching extracurricular activities anymore - they were guiding students through an enactive playground where AI and human creativity intertwined, creating learning experiences that were rich, dynamic, and full of unexpected joys.

This chapter in the school's journey with embodied AI has shown that the real power of technology in education lies not in what it can do for students but in how it can empower students to do more than they ever thought possible. As they looked to the future, the educators of Westfield High were excited to see what new innovative adventures awaited in the ever-evolving landscape of AI-enhanced learning.

Chapter 7:

From Theory to Practice—

Implementing Enactive AI

Curricula

The sun filters through the windows of a quiet library where Dr. Eliza Bennett, a pioneer in enactive AI education, is leading a workshop for a group of eager educators. The topic of discussion is the practical implementation of enactive AI curricula in schools across the country. The air is charged with a mix of excitement and apprehension as the educators prepare to bridge the gap between theory and practice.

Learning by Doing: The Power of Enactive Learning

Dr. Bennett begins the workshop by emphasizing the foundational principle of enactive learning: learning by doing. "Enactive learning," she explains, "is not about passive absorption of information, but about active engagement with the world around us. When we combine body-mind with AI, we create powerful, adaptive learning environments that respond to and evolve with each student's actions and perceptive experiences."

To demonstrate this concept, Dr. Bennett invites the workshop participants to engage in a simple exercise. She asks them to close their eyes and imagine teaching a lesson on photosynthesis. Then, she introduces an AI-powered augmented reality (AR) simulation where participants can manipulate virtual plants, adjusting light, water, and CO_2 levels to see the effects on photosynthesis in real time.

The difference in engagement and understanding is palpable. As one participant notes, "I've taught photosynthesis for years, but experiencing it this way gives me a whole new perspective. I can see how this would make the concept so much more accessible and memorable for students."

Dr. Bennett then delves deeper into the neuroscience behind enactive learning. She explains how physical engagement activates multiple areas of the brain simultaneously, creating stronger neural connections and enhancing memory formation. "When we combine mind-body interactions with the adaptive capabilities of AI," she says, "we're not just teaching content; we're training the brain to learn more effectively."

She shares a recent study where students who learned about cellular biology through an AI-driven VR experience showed a 35% improvement in long-term retention compared to those who learned through traditional methods (Chien, Su, Wu, & Huang, 2019). "The key," Dr. Bennett emphasizes, "is that these students weren't just observing; they were actively participating or embodying the cellular processes."

A real-world example of this approach can be seen in a study conducted by researchers who used augmented reality (AR) to enhance botanical learning. In this study, students interacted with virtual plants in a real-world environment, adjusting variables like light and water to observe the effects on plant growth and photosynthesis. This hands-on, interactive method significantly improved students' understanding and retention of complex biological concepts (Ibáñez & Delgado-Kloos, 2018).

Step-By-Step Guide to Developing AI-Enriched Lesson Plans

With the power of enactive learning firmly established, Dr. Bennett moves on to the practical aspects of curriculum design. She outlines a step-by-step process she uses for creating AI-enriched lesson plans:

1. **Identify Learning Objectives**: Clearly define what students should understand or be able to do by the end of the lesson.

2. **Choose Appropriate AI Tools**: Select AI technologies that align with your learning objectives and can create interactive, immersive experiences.

3. **Design Enactive Experiences**: Create scenarios or simulations that allow students to actively engage with the subject matter.

4. **Incorporate Adaptive Elements**: Utilize AI's ability to adjust difficulty and content based on student performance and engagement.

5. **Plan for Reflection**: Include opportunities for students to reflect on their experiences and integrate learning with their own unique set of experiences.

6. **Prepare Assessment Strategies**: Design assessment methods that can evaluate not just knowledge but also skills and understanding demonstrated through interaction with the AI environment. These methods should also present open-ended challenges, fostering innovative thinking where students and teachers collaboratively explore new active learning strategies.

7. **Consider Ethical Implications**: Ensure that the use of AI respects student privacy and promotes inclusive learning practices.

Dr. Bennett emphasizes that this process is iterative. "Your first AI-enriched lesson plan won't be perfect," she says, "but each iteration will bring you closer to creating truly transformative learning experiences."

To illustrate this process, Dr. Bennett walks the participants through the development of an AI-enriched lesson plan for a high school economics class. The lesson focuses on supply and demand dynamics:

- **Learning Objective**: Students will understand how various factors influence supply and demand in a market economy.

- **AI Tool**: An AI-powered market simulation that can model complex economic interactions.

- **Enactive Experience**: Students run virtual companies, making decisions about production, pricing, and marketing. The AI simulates market responses to these decisions.

- **Adaptive Elements**: The AI adjusts market conditions based on student decisions and introduces unexpected events (e.g., natural disasters, technological breakthroughs) to challenge students' understanding.

- **Reflection**: After each simulated quarter, students analyze their company's performance and market trends, discussing strategies with peers.

- **Assessment**: The AI provides detailed analytics on each student's decision-making process, understanding of key concepts, and ability to adapt to changing market conditions.

- **Ethical Considerations**: The simulation uses fictional companies and anonymized data to protect student privacy.

Dr. Bennett then invites participants to draft their own AI-enriched lesson plans, offering guidance and feedback as they work, encouraging students to generate questions, develop rubrics, and propose novel strategies, potentially inspiring future lessons.

How Flow Pedagogy Encourages Enactive Learning

The discussion then turns to the concept of "flow" in learning—an embodied experience of deep engagement and optimal experience first suggested by Mihaly Csikszentmihalyi in 1975.

Dr. Bennett elaborates on the concept of optimal learning conditions. She describes how crafting educational experiences that carefully calibrate difficulty to individual ability while offering well-defined objectives and instantaneous responses creates an ideal atmosphere for immersive learning. The integration of AI, she notes, enables this delicate equilibrium to be tailored for each learner. The embodied AI can continually fine-tune the experience, fostering students toward an optimal zone of engagement where they are neither overwhelmed nor under-stimulated.

Dr. Bennett emphasizes that this adaptive approach is key to fostering "flow"—periods of deep concentration and enjoyment that are highly conducive to learning. By leveraging AI in this way, educators can create dynamic learning environments that respond in real time to each student's progress, keeping them consistently challenged at a level that matches their evolving skills.

She shares an example from a mathematics class where an AI system presents students with increasingly complex geometry problems in a virtual 3D space. As students solve each problem, the AI adjusts the difficulty in real time, keeping each student challenged but not overwhelmed.

"In embodied flow," Dr. Bennett continues, "learning becomes almost effortless. Students lose track of time, fully immersed in the task at hand. This is where deep, lasting enactive learning occurs."

Dr. Bennett then introduces the concept of "micro-flow" - brief periods of intense focus and engagement that can occur even in short learning activities. She demonstrates how AI can be used to create flow throughout a lesson, optimizing student engagement and enhancing learning experiences.

For instance, in a language learning app, the AI might present a rapid-fire series of vocabulary challenges, each lasting just a few seconds. The difficulty and content of these challenges adapt in real-time based on the student's performance, keeping the mind-body fully engaged.

"By stringing together these embodied flow opportunities," Dr. Bennett explains, "we can create sustained periods of deep engagement, even with topics that students might traditionally find challenging or boring."

Real-Life Story: Revolutionizing History Education With AI Reenactments

To illustrate the transformative potential of enactive AI curricula, let's visit history teacher Mr. Saunders, who has implemented an AI-driven historical reenactment program.

In Mr. Saunders' class, students don VR headsets to step into key moments in history. The AI creates detailed, historically accurate environments and populates them with virtual characters based on historical figures and ordinary people of the time.

For a lesson on the American Revolution, students find themselves on the streets of Boston in 1773. They can interact with virtual colonists, British soldiers, and key historical figures, each powered by AI that allows for natural language conversations based on extensive historical data.

"What's remarkable," Mr. Saunders reports, "is how this experience brings history to life for the students. They're not just memorizing dates and events; they're embodying a deep, experiential understanding of the historical context, the motivations of different groups, and the complex factors that led to revolution."

The program has led to a significant increase in student engagement and test scores. More importantly, Mr. Saunders notes, it has fostered a genuine passion for history among his students, with many pursuing independent research projects inspired by their virtual experiences.

Mr. Saunders shares a particularly powerful moment from his class: "We had a student, Maria, who had always struggled with history. She couldn't see the relevance of these past events to her life. But after experiencing life as a young woman in colonial Boston through our VR program, something clicked. She started drawing parallels between the challenges faced by women in that era and issues of gender equality today. It sparked a passion for women's history that has completely transformed her engagement with the subject."

The success of the program has led to its expansion to other subjects. The school is now developing similar AI-driven immersive experiences for literature classes, allowing students to step into the worlds of classic novels, and for science classes, enabling virtual field trips to distant ecosystems or the depths of space.

Educator's Log: Learning Around Curriculum Design and Student Feedback

As the workshop progresses, Dr. Bennett emphasizes the importance of co-learning in innovative curriculum design, encouraging participants to share how they've incorporated student feedback, input, and observations in implementing and refining their enactive AI curricula. One particularly insightful account comes from Ms. Rodriguez, a science teacher who has been experimenting with a VR-based biology lab.

Ms. Rodriguez's VR lab allows students to conduct experiments that would be impossible or too dangerous in a traditional school setting, such as observing cell division at a microscopic level or exploring the effects of climate change on ecosystems over centuries.

"The most surprising thing," Ms. Rodriguez shares, "has been the student feedback. They're not just excited about the 'cool' technology; they're asking deeper questions, making connections between different concepts, and showing a level of scientific thinking I hadn't seen before."

She goes on to describe how the immediate, visual feedback provided by the AI has been crucial in addressing misconceptions and reinforcing correct understanding. "When a student makes a hypothesis and can instantly see the results played out in the virtual environment, it creates powerful learning moments," she explains.

However, Ms. Rodriguez also notes challenges, particularly in designing curricula that balance these high-tech experiences with more traditional learning methods and in ensuring that all students, regardless of their comfort with technology, can benefit from these new approaches.

Ms. Rodriguez shares a strategy she's developed to address these challenges: "We've created 'tech buddy' pairs in the class, partnering students who are more comfortable with technology with those who are less so. This not only helps all students engage with the AI tools but also promotes peer learning and collaboration."

She also emphasizes the importance of scaffolding the introduction of AI tools: "We start with simple, guided interactions and gradually increase complexity as students become more comfortable. This approach has helped even our most tech-hesitant students embrace these new learning methods."

Assessment and Feedback

Dr. Bennett builds on Ms. Rodriguez's experiences to discuss innovative approaches to assessment in an AI-enriched educational landscape. She emphasizes that AI tools can provide real-time insight and personalized learning analytics, enabling educators to tailor their instruction to the needs of each student.

"Traditional assessments often measure what a student knows at a single point in time," Dr. Bennett explains. "With AI, we can assess continuously, tracking not just what students learn, but how they learn, where they struggle, and how they apply knowledge in different contexts."

She demonstrates an AI system that analyzes student interactions in virtual environments, providing teachers with detailed insights into each student's problem-solving approaches, conceptual understanding, and skill development over time.

Dr. Bennett then introduces the concept of "stealth assessment"—assessment that occurs naturally as part of the learning process without students feeling like they're being tested. "In our AI-enriched environments," she explains, "every interaction becomes a potential data point. The AI can analyze patterns in these interactions to build a comprehensive picture of each student's learning journey."

She shares an example from a literature class where students explore a virtual world based on a novel they're studying. As they interact with characters and objects in this world, the AI assesses their understanding of plot points, character motivations, and themes—all without the students ever sitting down for a formal test.

"This approach not only provides richer data for teachers," Dr. Bennett notes, "but it also reduces test anxiety and allows students to demonstrate their embodied experiences in more natural, contextual ways."

Ethical Considerations

As the workshop draws to a close, Dr. Bennett leads a thoughtful dialogue on the ethical considerations of implementing embodied AI in education. The participants discuss the importance of data privacy, the potential for bias in AI algorithms, and the need for inclusive educational practices.

"As we embrace these powerful new technologies," Dr. Bennett cautions, "we must remain vigilant about their ethical implications. We need to ensure that our use of AI in education respects student privacy, promotes equity, and supports the holistic development of every learner."

The educators brainstorm strategies for addressing these concerns, from implementing strict data protection policies to regularly auditing AI systems for bias. They also emphasize the importance of involving diverse stakeholders in the curriculum development process. These stakeholders include:

- **Students**: The primary beneficiaries of the curriculum whose perspectives and experiences are crucial

- **Parents and Guardians**: To ensure the curriculum aligns with family values and expectations

- **Teachers From Various Disciplines**: To provide insights on subject-specific needs and cross-curricular opportunities

- **School Administrators**: To align the curriculum with broader educational goals and policies

- **AI and Technology Experts**: To offer technical insights and keep the curriculum current with technological advancements

- **Ethicists and Privacy Advocates**: To address ethical concerns and ensure responsible AI use

- **Special Education Specialists**: To ensure the curriculum is inclusive and accessible to all learners

- **Community Leaders**: To connect the curriculum with local contexts and needs

- **Industry Representatives**: To provide perspectives on future workforce needs

- **Educational Researchers**: To incorporate evidence-based practices and evaluate outcomes

By involving this diverse group of stakeholders, the educators aim to create a more robust, ethical, and inclusive AI-enhanced curriculum that addresses the needs and concerns of the entire educational community.

Dr. Bennett introduces the concept of "ethical AI literacy"—the idea that students should not only learn with AI but also about AI, including its ethical implications (Druga, 2019, Long & Magerko, 2020). "We need to prepare our students to be informed, critical users of AI technology," she explains. "This includes understanding issues of data privacy, algorithmic bias, and the broader societal impacts of AI."

She proposes integrating these topics into the curriculum, perhaps through a series of "AI ethics challenges," where students grapple with real-world ethical dilemmas related to AI in various contexts.

The group also discusses the importance of maintaining human connection in AI-enhanced learning environments. "While AI can provide personalized instruction and feedback," one participant notes, "we mustn't lose sight of the vital role of human teachers in providing support, motivation, and mentorship." AI can save time for teachers, allowing them to provide individualized feedback through written comments or one-on-one conversations, thereby gaining deeper insight into students' learning perceptions and goals.

As the workshop concludes, there's a palpable sense of both excitement and responsibility among the participants. They leave with practical strategies for implementing enactive AI curricula, a deeper understanding of the transformative potential of this approach, and a keen awareness of the ethical considerations that must guide their work.

Dr. Bennett's final words resonate with the group: "We stand at the threshold of a new era in education. By thoughtfully implementing enactive AI curricula, we have the opportunity to create learning experiences that are more engaging, more effective, and more equitable than ever before. The journey ahead is challenging, but the potential to positively impact countless learners makes it a journey well worth taking and opens exciting possibilities."

As the educators file out of the library, their minds are buzzing with ideas and possibilities. They know that implementing these new approaches will require hard work, creativity, and a willingness to embrace change. But they also know that they're part of a movement that has the potential to revolutionize education, opening up new worlds of learning for their students, yet well preparing their students for lifelong learning of embodying AI experiences.

In the coming weeks and months, these educators will return to their schools armed with new knowledge and inspired by the possibilities of enactive AI curricula. They'll face challenges and setbacks but also moments of triumph as they see their students engage with learning in new and profound ways. And as they continue to iterate and improve their approaches, they'll be at the forefront of a new chapter in the history of education—one where technology and human experience come together to unlock the full potential of lifelong learning and teaching.

Chapter 8:

The Global Classroom—AI and

Cultural Interaction

As the sun rises over one part of the world and sets in another, students from different continents gather in a shared virtual space, ready to embark on a journey of cross-cultural learning and collaboration. This is the global classroom, where AI serves as a bridge between languages, cultures, and geographical boundaries.

AI as a Tool for Global Education and Collaboration

The integration of AI in global education has created seamless, interactive learning environments that transcend geographical limitations. These AI systems serve multiple functions, revolutionizing how students from different parts of the world interact and learn together.

AI-powered real-time translation enables students to communicate in their native languages, breaking down language barriers that once hindered international collaboration. The embodied AI doesn't just translate words; it provides cultural context, explaining idioms, customs, and historical references that might be unfamiliar to students from different backgrounds.

Virtual environment creation is another key feature of these embodied AI systems. They generate immersive, culturally rich digital spaces for students to explore together, allowing for a more engaging and interactive learning experience. These environments can simulate historical events, ecological systems, or even futuristic scenarios, providing a platform for students to engage with complex concepts in a tangible way.

The embodied AI also acts as a collaboration facilitator, suggesting project groupings based on students' interests, skills, and learning goals. This fosters meaningful cross-cultural collaborations and ensures that students are challenged and engaged in ways that suit their individual learning styles.

Furthermore, by analyzing interactions and learning experiences, embodied AI helps educators develop culturally inclusive, globally relevant lesson plans. This adaptive curriculum generation ensures that the enactive learning material remains relevant, engaging, and sensitive to the diverse backgrounds of the students.

The Evolution of AI in Global Education

The journey to this sophisticated AI-driven global classroom was not without challenges. Language barriers, time zone differences, and technological limitations often hindered early attempts at cross-cultural digital collaboration. However, persistent efforts in educational technology have led to significant breakthroughs.

The focus shifted from solving basic translation problems to creating systems that could understand and convey cultural context. This led to the integration of advanced natural language processing and machine learning algorithms trained on vast datasets of cultural information, including literature, history, current events, and social norms from various cultures.

The development of immersive, shared virtual environments was another crucial step. These systems now use a combination of real-world data, user input, and procedural generation to create detailed, interactive virtual spaces that represent different cultural and geographical settings.

Cross-Cultural Learning in Action

In a typical Global Classroom session, students might find themselves exploring a virtual recreation of a significant geographical or historical location. For instance, in a lesson on water conservation, students could be transported to a digital representation of a water-stressed region.

As students from different parts of the world share their local perspectives and experiences, the embodied AI experience provides seamless translation and offers additional context. It might generate visual aids or infographics to illustrate complex concepts, making the learning experience more engaging and comprehensible for all participants.

The AI monitors engagement levels throughout the session, adjusting the pace and content to ensure all students are actively participating and learning. It identifies and highlights connections between the students' diverse perspectives, fostering a deeper understanding of the global nature of the challenges being discussed.

The Impact of Embodied Virtual Experiences

The power of these AI-facilitated virtual experiences lies in their ability to create mind-body-world learning experiences and learning opportunities. When students can virtually "walk" through different environments or scenarios, they're not just seeing or hearing about situations—they're experiencing them in a way that engages multiple senses and creates a sense of presence.

This type of embodied, experiential learning has been shown to increase empathetic perception, improve knowledge retention, and foster a deeper understanding of complex issues. Embodied AI's role in these experiences goes beyond just creating virtual mind-body-world environments. It actively shapes the enactive learning experience based on students' reactions and interactions, providing additional information or prompting questions to deepen exploration based on individual engagement.

Unleashing Creativity: Solving Worldwide Problems Through Cultural Diversity

The global classroom approach often involves collaborative projects that require students to apply their diverse perspectives to global challenges. For instance, students might be tasked with designing solutions to environmental issues that could be implemented in multiple global contexts.

In these projects, the embodied AI experience provides relevant data, suggests connections between different ideas, and offers real-time feedback to help students embody their concepts. It might introduce simulated challenges or future scenarios, encouraging students to think adaptively and consider long-term sustainability.

The resulting designs often demonstrate the power of cross-cultural collaboration, combining high-tech solutions with traditional practices or adapting ideas from one cultural context to solve problems in another. This approach not only enhances problem-solving skills but also fosters innovation through cultural synergy.

Fostering Global Competence

The impact of these cross-cultural experiences extends beyond the specific topics covered in each lesson. Students consistently report increased mind-body-world perceptual knowing, improved communication skills, and a greater appreciation for diverse perspectives. Many form lasting connections with their international peers, continuing their cultural exchange outside of formal learning sessions.

These experiences are particularly valuable in preparing students for an increasingly globalized future. The skills developed—cross-cultural communication, collaborative problem-solving, adaptability, and the ability to leverage AI tools effectively—are increasingly valued in the global workforce.

Challenges and Ethical Considerations

While the potential of AI in global education is immense, it also presents significant challenges and ethical considerations. Data privacy and security are primary concerns, particularly when dealing with minors and collecting data across international borders. Implementing robust systems to ensure that student data is protected, used ethically, and complies with various international data protection regulations is crucial.

Cultural sensitivity and bias prevention in AI systems is another critical consideration. Ensuring that AI doesn't perpetuate cultural biases or stereotypes requires ongoing effort, including diverse representation in development teams and continuous review and refinement of the AI's cultural knowledge base.

The digital divide presents another challenge. Not all schools have access to the high-speed internet and advanced hardware required for fully immersive experiences. Developing accessible versions of programs for schools with varying levels of technological infrastructure is an ongoing priority.

Expanding Beyond the Classroom

The success of AI-facilitated global learning has inspired initiatives to extend its impact beyond the school environment. Programs involving families in cross-cultural exchanges and collaborative community projects that address shared global challenges are beginning to emerge. These initiatives are expanding the reach of embodying world education, creating opportunities for entire communities to benefit from cross-cultural exchange of ideas and solutions.

The Neuroscience of Embodied Cross-Cultural Learning

Neuroscientific research is providing intriguing insights into the cognitive effects of embodied, cross-cultural learning experiences. Studies have shown increased activation in areas of the brain associated with empathy, complex problem-solving, and cognitive flexibility among students engaged in these programs (Immordino-Yang et al., 2019). Research indicates that such experiences can enhance neuroplasticity, particularly in regions of the brain linked to social cognition and emotional processing, such as the medial prefrontal cortex and the temporoparietal junction (Van den Bos et al., 2018). Additionally, some studies suggest structural changes in regions associated with language processing and social cognition after prolonged participation in such enactive learning experiences (Pulvermüller, 2018). These findings support the idea that embodied and cross-cultural learning can have lasting effects on the brain, fostering cognitive and emotional growth that goes beyond traditional learning methods.

The combination of embodied virtual experiences and real-time cross-cultural interaction appears to create a particularly effective form of learning, potentially building new neural pathways that bridge different cultural frameworks. These findings have implications not just for education but also for areas like conflict resolution and international cooperation.

AI as a Cultural Preservation Tool

Unexpected possibilities of embodied AI global learning have been its potential as a tool for cultural preservation. As students share their cultural knowledge and experiences, AI systems can build vast, dynamic databases of cultural information. Unlike traditional anthropological records, these systems capture culture as it's lived and experienced by young people today, creating a living, evolving, enacting global cultural knowledge.

Preparing for a Multilingual, Multicultural World

As the world becomes increasingly interconnected, the ability to navigate multiple languages and cultures is becoming ever more valuable. AI-facilitated global enactive learning is at the forefront of preparing students for this multilingual, multicultural future. Students are not just learning multiple languages but also developing the ability to switch between different cultural frameworks—a skill linguists call "cultural frame-switching."

The ability to seamlessly navigate different cultural contexts and adapt to different worldviews is likely to be an increasingly valuable skill in future job markets, particularly as companies build teams that span multiple countries and cultures.

The Role of Emotional Intelligence in Cross-Cultural AI Interactions

The role of emotional intelligence in AI-facilitated cross-cultural interactions is an emerging area of research (Kaur & Sharma, 2021). Efforts are being made to enhance AI systems' ability to recognize and respond appropriately to a wide range of culturally specific emotional expressions. When AI can accurately detect and respond to students' emotions, it can significantly enhance the quality of cross-cultural communication, providing additional context or suggesting different approaches to conversations when students appear frustrated or confused.

Looking to the Future: The Next Generation of Global AI Education

As AI-enhanced global education continues to evolve, new possibilities are emerging. Researchers are exploring the potential of brain-computer interfaces to create even more immersive and personalized learning experiences. There are plans for large-scale, AI-facilitated global youth summits, bringing together thousands of students from around the world to collaborate on addressing global challenges.

The principles of AI-facilitated global learning are also being considered for lifelong learning applications. AI-powered platforms could provide ongoing opportunities for cross-cultural learning and collaboration throughout a person's life, potentially revolutionizing areas like professional development, adult education, and even international relations.

Conclusion: Embracing a New Paradigm of Global Education

As we conclude our exploration of embodied AI in global education, it's clear that we stand at the threshold of a new era of embodied enactive learning. The integration of embodied AI, virtual reality, and cross-cultural collaboration is not just changing how we teach and learn; it's reshaping our understanding of how embodied AI education can and should be in a globally connected world.

This approach to education goes beyond the transmission of knowledge to the cultivation of global citizens—individuals who can think critically, collaborate across cultures, and contribute meaningfully to solving global challenges. As embodied AI experiences continue to evolve and our world becomes increasingly interconnected, these educational approaches will become ever more relevant and essential.

Looking to the future, we can envision a world where every learner, regardless of their location or background, has the opportunity to engage in rich, meaningful, transformative cross-cultural experiences. A world where embodied AI education potential is not confined by geographical or cultural boundaries but is instead a gateway to global understanding and collaboration. This is the promise of AI-enhanced global education—a promise that is beginning to be realized in classrooms around the world.

Chapter 9:

AI Continued Learning—

Professional Development and

Employment

In the rapidly evolving landscape of education and employment, one thing remains constant: learning never stops. As artificial intelligence continues to reshape our world, it's becoming increasingly clear that successful AI-driven careers require a unique set of skills. These skills go beyond mere technical knowledge; they encompass the ability to leverage AI tools effectively while maintaining the irreplaceable human expertise that drives innovation and critical thinking.

Embodied AI-enactive learning is not just transforming the educational experiences of students; it's revolutionizing professional development for educators and reshaping the very nature of employment across industries. This chapter explores how AI is changing the face of continued learning, from teacher training to adult education and beyond.

Preparing Educators for the AI Classroom

As AI becomes more prevalent in educational settings, it's crucial that educators are well-prepared to navigate and leverage these new tools. Teacher training programs are evolving to include AI literacy as a core component, ensuring that future educators are not just comfortable with AI but adept at integrating it into their teaching practices.

Universities are adapting their education programming to include learning on AI in education. For instance, Stanford University's Graduate School of Education offers various initiatives and discussions on the impact of AI in education, such as the AI+Education Summit, which explores how AI can be used to advance human learning while addressing ethical considerations (Stanford Graduate School of Education, 2023).

Similarly, the Harvard Extension School offers a Graduate Certificate in Artificial Intelligence. This program includes courses on machine learning, natural language processing, and the ethical and legal considerations of AI (Harvard Extension School, 2023). These programs are designed to equip educators and professionals with the knowledge and skills they need to thrive in AI-enhanced learning environments.

Embodied Teacher Training Approaches

Just as embodied AI enactive learning is proving effective for students, similar approaches are being adopted in teacher training. These methods go beyond theoretical understanding, allowing teachers to experience firsthand how AI can be integrated into their classrooms.

One innovative approach is the use of AI-powered virtual classrooms for teacher training. These simulations allow trainee teachers to practice their skills in a safe, controlled environment. The embodied AI experiences can simulate various student behaviors and learning needs, providing teachers with perceptual insight in addressing diverse classroom situations.

For example, the University of Central Florida has developed TeachLivE, a mixed-reality classroom simulator that uses avatar students. This system combines human control and programmed responses to create realistic classroom scenarios for teacher training

(University of Central Florida, 2023). While not strictly AI, this system points towards future possibilities where AI could power even more responsive and realistic training environments.

Stories of Teacher AI Education Journeys

The journey of integrating AI into teaching practices is unique for every educator. Let's explore a few stories that illustrate the diverse paths teachers are taking in their AI education journeys.

Maria Rodriguez, a high school science teacher with 15 years of experience, initially felt overwhelmed by the prospect of integrating AI into her classroom. She started her journey with online courses from platforms like Coursera and edX, which offered introductory courses on AI in education. Gradually, she began experimenting with AI-powered tools in her classroom, starting with a simple AI-driven question-answering system to support student inquiries.

On the other hand, James Chen, a fresh graduate from a teacher training program, entered the profession with AI integration as a core part of his skill set. His university program included courses on educational technology and AI, preparing him to use tools like AI-powered personalized learning platforms from day one of his teaching career.

Sarah Mbeki, a middle school math teacher in rural Kenya, faced unique challenges in her AI education journey. With limited access to technology, she relied on mobile-based AI tools and participated in distance learning programs offered by international universities. Despite the challenges, she has successfully integrated AI-powered math learning apps into her teaching, significantly improving student engagement and performance.

These stories highlight the diverse paths educators are taking to prepare for AI-enhanced teaching, underscoring the importance of flexible, accessible professional development opportunities.

Intelligent Tutoring Systems

Intelligent tutoring systems (ITS) represent one of the most promising applications of AI in education. These systems use AI to provide personalized instruction and feedback, adapting to each student's unique learning pace and style.

Carnegie Learning's MATHia platform, mentioned in Chapter 3, is a prime example of an ITS in action. This AI-powered math learning tool provides step-by-step guidance, adjusting the difficulty and approach based on each student's performance. It's being used in schools across the United States, with studies showing significant improvements in student learning outcomes (Institute of Education Sciences, What Works Clearinghouse, 2010).

For educators, understanding how to effectively integrate ITS into their teaching practice is becoming an essential skill. Professional development programs are increasingly focusing on how teachers can use the data and insights provided by these systems to inform their instruction and provide targeted support to students.

Assessment and Evaluation in the AI Classroom

AI is also transforming how we approach assessment and evaluation in education. Automated grading systems are becoming more sophisticated, capable of assessing not just multiple-choice questions but also essays and open-ended responses.

For instance, Gradescope, a platform developed by AI researchers at UC Berkeley, uses AI to streamline the grading process for written assignments and exams. It can automatically group similar answers together, allowing instructors to grade more consistently and efficiently (NVIDIA Developer, 2023).

However, the use of AI in assessment raises important questions about fairness, transparency, and the role of human judgment. Educators need to be trained not just in how to use these tools but in how to critically evaluate their outputs and ensure they're used ethically and effectively.

AI in Continued Education and Adult Learning

AI is not just transforming K-12 and higher education; it's also revolutionizing continued education and adult learning. Community colleges and adult education programs are increasingly leveraging AI to provide more flexible, personalized learning experiences.

For example, Southern New Hampshire University (SNHU) uses AI to enhance student support through initiatives like their AI chatbot, "Penny," which helps create personalized learning experiences for students (Southern New Hampshire University, 2023). While not specifically named IQUAL, these efforts point toward the broader trend of using AI to tailor educational paths for adult learners.

AI-powered platforms like Coursera and edX are also transforming how adults engage in continued learning. These platforms use AI to recommend courses, personalize learning experiences, and even provide career guidance based on a learner's progress and goals (Coursera, 2023; edX, 2023).

AI in Employment

The impact of AI on employment extends far beyond the education sector. As AI continues to automate routine tasks across industries, the nature of work is evolving. This shift is creating new demands on our education system to prepare students for an AI-driven workforce.

LinkedIn's Economic Graph project uses AI to analyze labor market trends, providing insights into emerging skills and job opportunities. This kind of AI-driven labor market intelligence is becoming invaluable for educators and career counselors in guiding students toward promising career paths (LinkedIn, 2023).

Moreover, AI is changing how companies approach hiring and professional development. AI-powered tools like HireVue use machine learning to analyze job candidate interviews. For example, HireVue's platform can assess candidates' responses, body language, and tone to provide a comprehensive evaluation (HireVue, 2024). Additionally, platforms like Degreed use AI to create personalized learning and development plans for employees. Degreed's AI-driven system, Maestro, helps identify skill gaps and recommends tailored learning paths to enhance employee development (Degreed, 2023).

Deeper Dive: AI-Enhanced Professional Learning Communities

Professional learning communities (PLCs) have long been a cornerstone of educator professional development. AI is now transforming these collaborative spaces, creating what we might call AI-enhanced professional learning communities (AEPLCs).

AEPLCs leverage AI to facilitate more effective collaboration and learning among educators. For instance, AI-powered analytics can identify patterns in student performance data across different classrooms, helping teachers and professors identify best practices and areas for improvement (McKinsey & Company, 2020). AI can also suggest relevant research articles, teaching resources, and professional development opportunities based on the specific challenges and goals discussed within the PLC (Edutopia, 2023).

One innovative application of AEPLCs is the use of AI-facilitated virtual reality spaces for teacher and professor collaboration. Imagine a virtual staffroom where educators from around the world can gather, sharing holographic representations of their classroom data, lesson plans, and student work. AI avatars could facilitate these discussions, providing real-time translation for international collaborations and suggesting connections between different educators' experiences and strategies (Harvard Graduate School of Education, 2024).

Gamification and AI in Professional Development

Gamification, combined with AI, could be a powerful tool for engaging educators in professional development. AI can personalize gamified learning experiences, adapting challenges and rewards to each educator's interests, teaching style, and areas for growth.

For example, a gamified AI system for professional development might challenge teachers to design and implement innovative lesson plans using AI tools. The system could provide virtual "missions" tailored to each teacher's subject area and student demographics. As teachers complete these missions, they earn badges and level up, unlocking new resources and challenges.

The AI could analyze the outcomes of these implemented lesson plans, providing feedback not just on the design but on the real-world impact on student engagement and learning outcomes. This creates a feedback loop that continually improves both the teacher's skills and the AI's understanding of effective teaching strategies.

AI and Micro-Credentials in Teacher Professional Development

Micro-credentials are gaining traction as a way to recognize specific skills and competencies, and AI is playing a crucial role in this trend.

AI-powered platforms can help educators identify relevant micro-credentials based on their current skills and career goals. These platforms can then provide personalized learning paths to achieve these micro-credentials, adapting the content and pace to the educator's progress.

For instance, Digital Promise, in partnership with AI education companies, is developing a system of AI-related micro-credentials for educators. These cover skills ranging from "Understanding AI Basics" to "Implementing AI-Enhanced Project-Based Learning." The AI platform not only guides teachers through the learning process but also assesses their submissions for the micro-credential, providing detailed feedback and suggestions for improvement.

Exam Proctoring With AI: Opportunities and Ethical Considerations

As online and distance learning become more prevalent, AI-powered exam proctoring systems are gaining popularity. These systems use a combination of AI technologies, including computer vision and machine learning, to monitor students during online exams, flagging potential instances of cheating.

For example, Proctorio, a widely used AI proctoring system, can detect unusual eye movements, background noises, or the presence of other people in the room (Proctorio, 2023). However, the use of such systems raises significant ethical concerns, particularly around privacy and equity.

Educators need to be well-versed in both the capabilities and limitations of these systems. Professional development in this area should cover not just the technical aspects of using AI proctoring tools, but also the ethical implications and best practices for ensuring fairness and protecting student privacy.

AI in Special Education: Tailored Professional Development

The application of AI in special education presents unique challenges and opportunities, necessitating specialized professional development for educators in this field. AI can provide powerful tools for personalized learning and communication support for students with special needs, but effectively leveraging these tools requires specific training.

For instance, AI-powered augmentative and alternative communication (AAC) devices are becoming increasingly sophisticated, using machine learning to predict and suggest words and phrases based on a user's patterns and context. Special education teachers need training not just in how to operate these devices but in how to effectively integrate them into their teaching strategies and how to train students and families in their use (Valencia et al., 2023).

Moreover, AI can assist in the early identification of learning disabilities and in tracking the progress of interventions. Professional development in this area might cover how to interpret AI-generated insights about student performance and behavior and how to use this information to refine individual education plans (IEPs).

AI and Cultural Competence in Education

As classrooms become increasingly diverse, cultural competence is more important than ever for educators. AI can play a significant role in developing this competence, both through personalized training for teachers and by providing real-time support in cross-cultural interactions.

AI-powered virtual reality simulations can place teachers in a variety of culturally diverse classroom scenarios, allowing them to practice culturally responsive teaching strategies in a safe, low-stakes environment. These simulations can adapt based on the teacher's responses, providing increasingly complex scenarios as the teacher's skills improve.

In real classroom settings, AI language models can provide real-time suggestions for culturally appropriate language and teaching strategies. For example, an AI system might alert a teacher to the cultural connotations of a particular example they're using or suggest alternative explanations that might resonate better with students from different cultural backgrounds.

The Gig Economy and AI: Implications for Educators and Students

The rise of the gig economy, temporary or independent jobs facilitated by AI-powered platforms, is changing the employment landscape. This shift has significant implications for both educators and students.

For educators, the gig economy presents both challenges and opportunities. On one hand, it may lead to more precarious employment in traditional teaching roles. On the other, it opens up new possibilities for educators to leverage their skills in diverse ways, such as creating and selling online courses, offering tutoring services through AI-matched platforms, or providing consulting services to edtech companies.

For students, preparation for the gig economy needs to be integrated into career education. This includes developing personal branding and financial management skills and leveraging AI tools to find and manage gig work. Here are some examples of how AI can support this preparation:

- **Personal Branding**

  - AI-powered writing assistants like Grammarly or GPT-based tools can help students craft compelling personal statements and optimize their online profiles.

  - Image recognition AI can analyze profile pictures and suggest improvements for more professional presentation on platforms like LinkedIn.

- **Financial Management**

  - AI-driven budgeting apps like Mint or YNAB can teach students how to manage irregular income streams common in gig work.

  - Predictive analytics tools can help students forecast income based on gig economy trends, aiding in financial planning.

- **Finding Gig Work**

  - Platforms like Upwork and Fiverr use AI algorithms to match freelancers with suitable projects based on their skills and experience.

  - AI chatbots on job search sites can guide students through the process of finding and applying for gig work opportunities.

- **Managing Gig Work**

  - Project management tools with AI capabilities, such as Trello or Asana, can help students juggle multiple gigs efficiently.

  - AI-powered time-tracking tools like RescueTime can help students understand their productivity patterns and optimize their work schedules.

By integrating these AI tools and approaches into career education, schools can help students develop the adaptability and entrepreneurial skills needed to thrive in the evolving job market, where gig work is becoming increasingly prevalent.

AI and Entrepreneurship in Education

The intersection of AI and education is creating fertile ground for entrepreneurship. Educators with an entrepreneurial spirit have unprecedented opportunities to innovate and create new solutions for longstanding educational challenges.

For instance, an elementary school teacher might develop an AI-powered app that gamifies vocabulary learning, adapting to each child's interests and learning style. A high school physics teacher might create a virtual reality platform that uses AI to generate personalized physics problem scenarios based on each student's real-world interests.

Professional development programs are beginning to incorporate modules on "eduprenuership," teaching educators how to identify needs in the education market, develop AI-enhanced solutions, and navigate the business side of bringing an educational product to market.

Global Perspectives on AI in Education and Employment

As we consider the impact of AI on education and employment, it's crucial to maintain a global perspective. The challenges and opportunities presented by AI vary significantly across different countries and cultures.

In some developing countries, AI is seen as a potential solution to teacher shortages and lack of access to quality educational resources. For example, the African app Eneza Education uses AI to provide personalized learning content via text message, making education accessible even in areas with limited internet connectivity (Eneza Education, 2024).

In countries with aging populations, such as Japan, AI is being explored as a way to support lifelong learning and career transitions for older adults. AI-powered platforms are being developed to help seniors acquire new skills and find flexible employment opportunities that match their experience and abilities (World Economic Forum, 2021).

Understanding these global variations is crucial for educators and policymakers as they prepare students for an increasingly interconnected, AI-driven global economy.

The Role of Emotional Intelligence in an AI-Driven World

Before we delve into the role of emotional intelligence in an AI-driven world, it's crucial to understand what we mean by this term in the context of learning and embodied AI cognition experiences. In this framework, emotional intelligence refers to the ability to recognize, understand, and manage our own emotions as well as to recognize, understand, and influence the emotions of others (Goleman, 1995).

In the realm of education and AI, emotional intelligence encompasses several key aspects:

- **Self-Awareness**: The ability to recognize one's own emotional states and how they impact learning and decision-making (Goleman, 1995)

- **Self-Regulation**: Managing emotions and impulses, especially in challenging learning situations (Goleman, 1995)

- **Motivation**: Using emotions to drive persistence and engagement in learning tasks (Goleman, 1995)

- **Empathy**: Understanding and responding to the emotions of peers and instructors in collaborative learning environments (Goleman, 1995)

- **Social Skills**: Navigating complex social interactions in both physical and AI-enhanced virtual learning spaces (Goleman, 1995)

For example, in an AI-enhanced classroom, emotional intelligence might manifest as:

- A student recognizing their frustration with a difficult concept and using that awareness to seek help from an AI tutor or human teacher (Forbes, 2024)

- Learners effectively collaborating on a virtual reality project, reading and responding to each other's emotional cues even in a digital space (Forbes, 2024)

- A student empathizing with historical figures in an AI-simulated historical scenario, leading to deeper understanding and engagement with the material (Forbes, 2024)

As we prepare students for future employment, we need to emphasize that emotional intelligence is not in competition with AI but a crucial complement to it. The most successful professionals in an AI-driven world will be those who can effectively combine the analytical power of AI with these uniquely human capabilities of empathy, ethical reasoning, and complex social interaction (Oxford Group, 2024).

For instance, in a future workplace, an employee might use AI to analyze vast amounts of data but then apply emotional intelligence to present those findings in a way that resonates with colleagues and stakeholders, navigating the human elements of decision-making and change management (Oxford Group, 2024).

By introducing and nurturing these emotional intelligence skills alongside AI literacy, we can prepare students for a future where human and artificial intelligence work in synergy, each enhancing the capabilities of the other (Oxford Group, 2024).

Future Trends

As we look to the future, several trends are emerging in the intersection of AI, education, and employment:

- **Hyper-Personalization**: AI will enable increasingly personalized learning experiences, not just in terms of content and pace but also in learning style, interests, and career goals. For instance, Carnegie Mellon University's AI-powered writing tutor, developed in collaboration with Turnitin, analyzes a student's writing style, strengths, and weaknesses. It then provides personalized feedback and suggests resources tailored to the student's specific needs, learning style, and even career aspirations (Carnegie Mellon University, 2020). A student interested in journalism, for example, might receive writing prompts and feedback geared toward news-style writing.

- **Lifelong Learning Portfolios**: AI could help create and manage comprehensive lifelong learning portfolios, tracking skills and knowledge gained through formal education, work experience, and informal learning. LinkedIn Learning exemplifies this trend, using AI to track a user's course completions, skills development, and work experiences. It then suggests new courses or certifications based on the user's career trajectory and emerging industry trends (LinkedIn Learning, 2019). This creates a dynamic, AI-curated professional portfolio that evolves with the learner's career.

- **AI Collaboration Skills**: The ability to effectively collaborate with AI systems will become a crucial skill across industries. Education will need to focus on developing these human-AI collaboration skills. Georgia Tech's "Jill Watson" AI teaching assistant project not only helps students with course-related queries but also teaches them how to effectively interact with

AI systems. Students learn to phrase questions in ways that AI can understand and interpret, preparing them for future work environments where human-AI collaboration will be commonplace (Goel & Polepeddi, 2018).

- **Ethical AI Education**: As AI becomes more prevalent, education on the ethical implications of AI will become increasingly important, both for those developing AI systems and for general digital citizenship. The University of Helsinki's free online course "Ethics of AI" is an example of this trend, covering the ethical implications of AI in various sectors. The course uses AI-powered simulations to present ethical dilemmas, allowing students to explore the consequences of their decisions in a safe environment (University of Helsinki, 2021). This type of education is becoming increasingly common in computer science and data science programs worldwide.

- **Augmented Intelligence in Education**: Rather than replacing human teachers, AI will increasingly be used to augment human intelligence, providing educators with powerful tools to enhance their teaching. Third Space Learning, a UK-based tutoring company, demonstrates this by using AI to support human tutors during one-on-one math sessions. The AI analyzes the conversation in real time, providing tutors with suggestions for explanations, examples, and teaching strategies based on the student's responses and learning patterns (Third Space Learning, 2019). This allows the human tutor to provide more effective, personalized instruction.

Implications for Stakeholders

This AI revolution in continued learning and employment has implications for all stakeholders in education:

- **Educators** need to embrace lifelong learning themselves, continuously updating their skills to effectively integrate AI into their teaching practices.

- **Students** must be prepared for a future where AI is ubiquitous, developing not just technical skills but also the critical thinking and creativity that will remain uniquely human domains.

- **Parents** should be aware of how AI is changing education and career landscapes, supporting their children in developing the skills needed for future success.

- **Administrators** need to make informed decisions about AI integration, balancing the potential benefits with considerations of equity, ethics, and efficacy.

- **Educational Businesses**, from textbook publishers to edtech startups, must adapt to this new landscape, developing AI-enhanced products and services that truly support teaching and learning.

- **Entrepreneurs** in the education space have unprecedented opportunities to innovate, creating new AI-powered solutions to address longstanding challenges in education, as well as small business environments worldwide.

Conclusion: Embracing Continuous Adaptation

As we navigate the ongoing AI revolution in education and employment, we must shift our focus from mere adaptation to embracing enactive learning. This approach goes beyond continuously adapting to change; it emphasizes an embodied flow pedagogy that engages learners in the present moment, allowing them to actively shape their learning environment.

Enactive learning, rooted in the work of Varela et al. (1991), views cognition as an embodied, action-oriented process. In the context of

AI-enhanced education, this means not just responding to technological changes but actively engaging with and shaping these technologies through our interactions.

The pace of technological change shows no signs of slowing, and the specific AI tools and applications we use today may be obsolete within a few years. Therefore, the goal of education in this AI age should not be to master any particular technology but to cultivate a mindset of enactive, embodied learning. This approach allows learners to flow with the currents of technological change, not just adapting to them but actively participating in their evolution.

We need to help students (and ourselves) become comfortable with uncertainty, approaching new technologies not just with curiosity but with a readiness to engage, experiment, and co-create. In this paradigm, change becomes more than an opportunity for growth—it becomes a canvas for innovation and discovery.

By embracing enactive learning and embodied flow pedagogy, we prepare learners not just to adapt to the future but to actively shape it. This approach recognizes that learning is not a passive process of absorbing information but an active, embodied experience of engaging with the world—a world increasingly infused with AI.

In this way, we move beyond continuous adaptation to a more dynamic, creative relationship with AI and technological change. Learners become not just consumers of technology but co-creators, ready to navigate and shape the exciting possibilities that emerge at the intersection of human creativity and artificial intelligence.

This adaptive mindset is crucial not just for navigating the changing landscape of work but for actively shaping it—an embodied or flow mindset or flow pedagogy. As AI continues to evolve, we have the opportunity—and the responsibility—to guide its development and application in ways that enhance human potential.

For educators and all learners throughout life, this means continually refining our understanding of how AI can be used to create more engaging, effective, and equitable learning experiences. It means being willing to experiment with new tools and approaches, learn from both successes and failures, and share these learnings with colleagues around the world.

For students, embracing continuous adaptation means developing a strong foundation of core skills—critical thinking, creativity, communication, and collaboration—while also cultivating the ability to quickly acquire new knowledge and skills as needed. It means learning how to learn, and how to embody AI tools as partners in the learning process.

For educational institutions, it means creating flexible, responsive systems that can quickly adapt to changing needs and opportunities. This might involve reimagining traditional degree programs, creating new interdisciplinary fields of study that bridge technology and human sciences, or developing innovative partnerships with industry to ensure that education remains relevant to the rapidly evolving world of work.

As we look to the future, it's clear that the integration of AI into education and employment will continue to accelerate. But this is not a future that's happening to us; it's one that we embody through our enactive learning experiences and practice. By embracing embodied AI as a tool for enhancing human capabilities, by thoughtfully addressing the ethical challenges it presents, and by committing to a path of continuous learning and adaptation, we can foster a future where technology while harmoniously unlocking new or unexpected realms of potential.

Chapter 10:

The Future Is Now—Preparing for the AI Revolution in Education

As we stand on the brink of a new era in education, the integration of artificial intelligence is not just a possibility—it's an unfolding reality. The AI revolution in education is reshaping how we teach, learn, and think about knowledge itself. This chapter explores the cutting edge of this revolution and peers into the future of embodying AI-enactive learning.

Predictions and Preparations for a Constantly Evolving Pedagogy

The rapid advancement of AI technology means that educational pedagogy must evolve at an unprecedented pace. Gone are the days when a teaching method could remain static for decades. Today's educators must be prepared for constant change, adapting their approaches as new AI tools and capabilities emerge. However, this is not new to the educators who have embraced enactive learning while incorporating a flow pedagogy that opens possibilities for all body-mind-world or unknown potential experiences in all "classrooms."

One of the most significant shifts is the move toward personalized learning at scale. AI systems are becoming increasingly adept at analyzing individual student data—learning styles, pace, strengths, and weaknesses—and tailoring educational content accordingly. This level of personalization, once a luxury available only through one-on-one tutoring, is becoming accessible to entire classrooms and school districts.

Predictive analytics in education is another area of rapid growth. AI systems can now analyze vast amounts of data to identify students at risk of falling behind or dropping out, allowing for early intervention. These systems don't just flag potential issues; they can suggest personalized intervention strategies based on what has worked for similar students in the past.

The role of the teacher is also evolving. Rather than being made obsolete by AI, teachers are becoming facilitators and guides, helping students navigate the wealth of information and tools at their disposal. Professional development for educators increasingly focuses on AI literacy—understanding how to effectively integrate AI tools into the classroom and how to teach students to interact with AI critically and ethically.

Curriculum design is another area undergoing significant change. Static, standardized curricula are giving way to dynamic, adaptive learning pathways. These AI-driven curricula can adjust in real time based on student performance and engagement, ensuring that each learner is consistently challenged at the appropriate level.

As we prepare for this constantly evolving educational landscape, flexibility and adaptability are key. Schools are reimagining their physical spaces and schedules to accommodate AI-enhanced learning. Many are adopting longer class periods, such as 120-minute blocks, which allow for a more diverse range of activities within a single session. These extended periods create flexible environments that can easily shift between individual work, small-group collaboration, and full-class activities.

In these reimagined classrooms, students might sit at tables, working on rubric-guided activities while teachers circulate, providing one-on-one assistance and discussing individual learning goals. AI tools can support this model by providing personalized learning pathways and real-time feedback, freeing teachers to focus on higher-level guidance and support. This approach allows for a blend of AI-assisted independent learning, peer collaboration, and targeted teacher intervention.

The shift to longer, more flexible class periods also raises important questions: How can AI tools best support extended learning sessions? How might the rhythm of a 120-minute class differ when AI is integrated? How can we balance screen time with hands-on activities in these longer blocks?

Professional development for educators is becoming an ongoing process, with continuous enactive learning opportunities to keep pace with both pedagogical and technological advancements. This includes not only mastering new AI tools but also learning how to effectively integrate them into longer, more fluid class structures.

School Transformation Through AI Integration

The transformation of education through AI integration is not a future prospect—it's happening now in schools around the world. One striking example is the comprehensive AI integration program implemented at Summit Public Schools in California, Piedmont City School District in Alabama, and Minerva Schools, a global higher education institution.

Three years ago, the Piedmont City School District was facing common challenges such as limited resources, a diverse student body, and the pressure to prepare students for an increasingly technology-driven world. Today, the district is at the forefront of AI-enhanced education (Project Pals, 2023). The transformation began with a comprehensive needs assessment and a commitment to gradual, thoughtful integration of AI tools. They started by implementing AI-powered personalized learning platforms in math and science. This system provided students with customized problem sets and real-time feedback, allowing teachers to focus on individual student needs rather than one-size-fits-all instruction (Project Pals, 2023).

For instance, in a 10th-grade algebra class, the AI system first assessed each student's current understanding of quadratic equations. Based on this assessment, it generated personalized problem sets for each student. As students worked through their customized problems, the AI provided immediate feedback, explaining errors and offering hints when students got stuck. The platform also adjusted the difficulty of problems based on each student's performance, ensuring they were appropriately challenged.

This personalized approach freed teachers from having to lecture the entire class on a single topic. Instead, teachers could monitor progress in real time, allowing them to identify students who needed additional support and provide targeted, one-on-one instruction (Project Pals, 2023). By the end of the semester, teachers reported significant improvements in student engagement and understanding.

Encouraged by early results, Summit Public Schools also expanded its AI integration. In the English department, AI-powered writing tools helped students improve their writing skills through instant, detailed feedback. AI-enabled tools have also been incorporated into history classes, allowing students to explore virtual environments and experience historical events, enhancing engagement and understanding of complex concepts (Stanford HAI, 2023).

One of the most dramatic improvements came from Minerva Schools, a higher education institution that adopted AI-driven platforms for continuous personalized feedback and assessments. The real-time feedback allowed students to move at their own pace, receive instant responses, and work closely with faculty members who used AI to enrich their instruction (Minerva University, 2023a; Minerva University, 2023b).

However, these transformations were not without challenges. Some teachers initially felt threatened by the new technology, fearing that it might make their roles obsolete. Schools addressed these concerns through comprehensive professional development programs, helping teachers understand how AI can enhance, rather than replace, their teaching practices (Stanford HAI, 2023). Additionally, data privacy was a significant concern. Schools worked with AI providers and cybersecurity experts to ensure robust protections for student data while also teaching students about digital literacy and data privacy (Minerva University, 2023a).

Three years into the program, the results are impressive. Test scores have improved across the board, with notable gains among previously struggling students. Teachers report feeling more effective and less burned out, and student engagement is at an all-time high. Perhaps most importantly, students are acquiring the technological literacy and critical thinking skills that will serve them well in an AI-driven future (Project Pals, 2023).

Educators' Vision: Experiential Learning for Future Challenges and Opportunities

As we look ahead to the future of education, many educators are focusing on experiential learning as a key strategy for preparing students for the challenges and opportunities of an AI-driven world. This approach, which emphasizes learning through embodied experience and reflection, is seen as crucial for developing the adaptability, creativity, and critical thinking skills that will be essential in the future job market.

One emerging trend is the use of AI-enhanced simulations and game-based learning. These immersive experiences allow students to engage with complex concepts in a hands-on way, making abstract ideas concrete and memorable. For instance, economics students might manage a simulated global economy, dealing with AI-generated crises and opportunities. This not only teaches economic principles but also hones decision-making skills and systemic thinking.

Project-based learning is also being reimagined with AI support. Students are increasingly working on long-term, interdisciplinary projects that mirror real-world challenges. AI tools assist in these projects by providing data analysis, suggesting resources, and even offering creative prompts to spark new ideas. This approach helps students develop collaboration skills, project management abilities, and the capacity to synthesize information from various sources—all crucial skills in the modern workplace.

Another area of focus is developing students' ability to work alongside AI systems effectively. This goes beyond mere technical proficiency to include understanding AI's capabilities and limitations, being able to critically evaluate AI-generated information, and knowing how to guide AI tools to produce desired outcomes. Some schools are introducing "AI apprenticeship" programs, where students work on real-world projects in partnership with AI systems, learning how to embody AI as a powerful collaborator.

Emotional intelligence and human-centric skills are also receiving increased attention. As AI takes over more routine cognitive tasks, uniquely human skills like listening skills, creative problem-solving, and complex communication are becoming more valuable. Educators are developing curricula that explicitly foster these skills, often using AI-powered tools to provide personalized coaching and feedback on interpersonal interactions.

Ethical considerations are being woven throughout these experiential learning approaches. Students are encouraged to grapple with the ethical implications of AI in various contexts, from privacy concerns to the potential societal impacts of automation. This ethical grounding is seen as crucial for developing responsible AI practitioners and informed citizens.

As we move forward, the goal of education is increasingly seen not as imparting a fixed body of knowledge but as developing adaptable, lifelong learners who can thrive in a rapidly changing world. Experiential learning, enhanced by AI, (enactive learning through embodied AI) is proving to be a powerful tool in achieving this goal, as well as providing opportunities for learners to transcend their perceived limitations and unlock previously untapped embodied potential.

Lifelong Learning in the Age of AI

The concept of lifelong learning is not new, but AI is revolutionizing what it means and how it's achieved. As the pace of technological change accelerates, the half-life of professional skills is shrinking, making continuous learning not just beneficial but essential for career success.

AI is making lifelong learning more accessible and effective than ever before. Adaptive learning platforms can now create personalized curricula for adult learners, taking into account their existing knowledge, learning style, and professional goals. These systems can identify skill gaps and suggest targeted learning experiences to address them, ensuring that unique learners are always focusing on the most relevant and impactful content.

Microlearning, facilitated by AI, is becoming increasingly popular. This approach breaks down learning into small, manageable chunks that can be easily integrated into busy adult lives. AI systems can deliver these bite-sized lessons at optimal times and in formats that suit individual learners, maximizing mind-body retention and application of new knowledge.

Virtual and augmented reality, powered by AI, are opening up new possibilities for immersive, experiential learning in professional contexts. For instance, surgeons can practice complex procedures in virtual environments, receiving real-time feedback and guidance from AI systems. This allows for safe, repeated practice of high-stakes tasks, accelerating skill development.

AI-powered career navigation tools are also emerging, helping individuals make informed decisions about their learning journeys. These systems can analyze labor market trends, an individual's skills and interests, and emerging technologies to suggest promising career paths and the learning experiences needed to pursue them.

Moreover, AI is enabling new forms of collaborative learning among professionals. Intelligent networking systems can connect learners with peers and mentors based on complementary skills and experiences, facilitating knowledge sharing and collaborative problem-solving across geographical and organizational boundaries.

As AI continues to evolve, the boundary between working and learning is blurring. AI assistants in the workplace can provide just-in-time learning, offering guidance and information as needed to complete tasks. This integration of learning into daily work helps ensure that new knowledge is immediately applied and retained or embodied through experience.

The future of lifelong learning, empowered by AI, is one of continuous, personalized skill development seamlessly integrated into both professional and personal life. It promises to help individuals stay relevant and adaptable in a rapidly changing world, turning the challenge of constant technological change into an opportunity for ongoing growth and development.

AI and the Unfolding of Human Potential: An Enactive Pedagogy

As AI systems become more sophisticated, there's growing interest in how they cannot just assist in learning but actively contribute to the unfolding of human potential. This approach, often referred to as enactive pedagogy, sees learning not as the passive reception of information but as an active, embodied process of engaging with the world.

In this context, AI is seen not just as a tool but as a partner in the learning process, helping to create rich, responsive environments that encourage exploration, creativity, and self-discovery. This partnership between human learners and AI systems is opening up new possibilities for innovation and personal growth.

One exciting development is the use of AI to foster creativity and innovation. AI systems can now generate novel ideas, combining concepts in unexpected ways. When students interact with these AI creativity tools, they're exposed to new possibilities and perspectives, challenging their assumptions and sparking innovative thinking. The goal is not for the AI to replace human creativity but to serve as a springboard for it, pushing students and teachers to think beyond their usual boundaries.

AI is also being used to create more immersive and interactive storytelling experiences. Students can now step into AI-generated narratives, making choices that shape the story's direction. This not only makes learning more engaging but also helps students understand the consequences of decisions and the complexity of narrative structures. In writing classes, AI writing partners can offer suggestions and alternative phrasings, helping students refine their voice and style.

In the sciences, AI is enabling students to engage with complex systems in new ways. For instance, AI-powered simulations allow students to manipulate variables in ecosystems or chemical reactions, seeing the results play out in real time. This hands-on, exploratory approach helps students develop a deep, intuitive understanding of scientific principles.

Perhaps most profoundly, AI is helping to create enactive learning experiences that adapt not just to a student's cognitive abilities, but to their dispositions, motivations, and personal goals. By analyzing facial expressions, tone of voice, and patterns of interaction, AI systems can gauge a student's readiness and level of engagement, adjusting the learning experience accordingly. This comprehensive responsiveness helps create a more supportive and effective learning environment.

As we look to the future, the potential of embodied AI in education goes far beyond mere efficiency or information delivery. It's about creating dynamic, responsive learning environments that encourage exploration, creativity, and self-discovery. It's about using technology not to replace human potential but to unlock it, helping each individual write their own unique story of growth and discovery.

In this vision of the future, education becomes a lifelong journey of self-actualization, with AI as a constant companion and collaborator. The classroom of tomorrow is not just a place to acquire knowledge but a launchpad for human potential, where each learner can chart their own course through the vast landscape of knowledge and experience.

As we stand on the brink of this AI-enabled educational revolution, the possibilities are both exciting and challenging. It will require us to rethink our assumptions about learning, intelligence, and human potential. But if we can harness the potential of AI while keeping human growth and flourishing at the center of our educational endeavors, we have the opportunity to unleash new realms of human achievement and understanding.

The future of education is not just about AI teaching humans but about humans and AI learning together, each augmenting the other's strengths. It's a future where technology and humanity converge to enact learning experiences that are more personalized, more engaging, and more transformative than ever before. As we step into this future, we're not just preparing for the AI revolution in education—we're actively shaping it, ensuring that it serves the highest aspirations of human learning potential.

Conclusion

As we conclude our exploration of embodied AI in education, we stand at the threshold of a new era in learning. This book has taken us on a journey through the transformative potential of AI in education, revealing how it can nurture curious minds, create immersive learning experiences, and prepare all stakeholders for a future where human creativity and artificial intelligence work in synergy.

Embodied AI education is not just about integrating technology into classrooms; it's about reimagining the very nature of learning. It's about creating educational experiences that engage an embodied consciousness—mind, body, and world—in the process of discovery and growth.

Key Insights and Takeaways

As we conclude our exploration of embodied AI education, let's return to where our journey began: that crisp autumn morning when young Alex discovered the magic of photosynthesis during a simple walk in the park.

Remember how Alex struggled with the abstract concept in the classroom, but his understanding blossomed as he placed his hand on the rough bark of that massive oak tree? This moment encapsulates the essence of embodied AI education and the key insights we've gathered along our journey.

As Alex looked up at the canopy of golden leaves, an AI-enhanced reality overlay showed him the invisible process of photosynthesis happening in real time. This wasn't AI replacing his teacher but partnering with her to create a powerful learning experience. It exemplifies how AI can enhance and amplify human capabilities in education rather than replace them.

The AI system recognized Alex's moment of discovery and adapted instantly, tailoring the complexity of the information to his sudden burst of curiosity. This personalization at scale is one of the most promising aspects of embodied AI in education, making it possible to cater to individual needs, interests, and learning styles in ways never before possible.

As Alex continued his walk, the AI engagement didn't stop at visual overlays. He could hear the soft rustle of leaves transforming carbon dioxide into oxygen, feel the vibration of water moving up the tree's trunk through his hand, and even smell the subtle changes in the air around him. This multi-sensory, immersive learning experience engaged Alex on multiple levels, making the lesson more engaging and memorable.

The AI system didn't confine Alex's learning to biology alone. It drew connections to the chemistry of photosynthesis, the physics of light, and even the role of trees in local ecosystems and global climate. In this way, AI facilitated interdisciplinary learning, breaking down the silos between subjects and helping Alex see connections across different areas of knowledge, actualizing real-world applications.

As Alex's walk came to an end, the AI system suggested ways he could continue exploring the concept of photosynthesis at home, in his neighborhood, and even in different seasons. This exemplifies how embodied AI education cultivates skills for lifelong learning, preparing learners for a future of continuous adaptation and growth.

Throughout this embodied AI-enhanced nature walk, Alex's data—his location, his interactions, his learning progress—was being collected and analyzed. This serves as a reminder of the ethical considerations we must keep in mind. As we embrace AI in education, we must remain vigilant about issues of privacy, equity, and the ethical use of technology.

Finally, imagine if Alex could share his discovery in real time with students on the other side of the world, perhaps comparing the photosynthesis process in his oak tree with that of a tropical plant in the Amazon rainforest. This represents the exciting possibilities AI opens for global collaboration and cross-cultural learning on a scale never before possible.

Alex's journey from confusion to comprehension, from abstract concept to embodied understanding, mirrors our own journey through this book. It reminds us that at the heart of embodied AI education is not the technology itself but the transformative learning experiences it can create when thoughtfully integrated with human guidance and the physical world around us.

A Call to Action for Embodied AI Education

The time for embodied AI education is now. We call on educators, students, parents, administrators, and entrepreneurs to embrace this new paradigm of learning:

- **Educators**: Commit to ongoing professional development in AI. Experiment with AI tools in your teaching and share your experiences with colleagues.

- **Students**: Approach AI with curiosity and critical thinking. Learn to work alongside AI tools while developing your uniquely human skills.

- **Parents**: Support your children's exploration of AI-enhanced learning while fostering a balanced approach to technology use.

- **Administrators**: Invest in AI infrastructure and professional development. Create policies that promote innovative, ethical use of AI in education.

- **Entrepreneurs**: Develop AI tools that truly enhance learning experiences, always keeping the needs of educators and students at the forefront.

Envisioning the Transformative Potential of AI in Schools

Imagine classrooms where virtual reality transports students to ancient civilizations or inside the human body. Envision AI tutors providing personalized support to every student, freeing teachers to focus on higher-order learning and socio-emotional development. Picture a world where language barriers dissolve, allowing students to collaborate with peers across the globe in real time.

This is the promise of embodied AI education. It's a future where learning is not confined to classrooms or textbooks but happens everywhere, all the time, tailored to each learner's needs and interests, an evolving curriculum.

How Will You Embody AI?

As we conclude, we invite you to reflect on your own journey with AI in education. How will you embody AI in your learning or teaching practice? What possibilities excite you? What challenges do you foresee?

Remember, the stories and examples shared in this book are just the beginning. They are invitations to a larger conversation about the future of education. Your experiences, your innovations, your questions—these are all part of the ongoing story of embodied AI education.

Looking Ahead: Implementing Embodied AI Education

As we look to the future, many questions arise about the practical implementation of embodied AI education:

- How can schools effectively integrate AI tools while ensuring equitable access for all students?

- What does professional development for Embodied AI Education look like in practice?

- How can we measure the success of Embodied AI Education initiatives?

- What ethical frameworks should guide our use of AI in educational settings?

- How can we foster a culture of innovation and continuous learning in our educational institutions?

These questions and more will be addressed in our upcoming implementation guide, which will provide practical guidance for bringing the concepts of embodied AI education to life in your school or learning environment.

The journey of embodied AI education continues. It's a journey of unleashing creative discovery through embodied human experience, nurturing minds and hearts, and empowering learners for positive lifelong learning. As we go on this journey together, let us move forward with curiosity, compassion, and a commitment to fostering a spirit of continuous questioning, seeking our greatest potential to be the best we can be.

As we conclude this exploration of embodied AI in education, we invite you to reflect on your own journey. Are you ready to challenge your assumptions about what education can be? How might embodied

AI change your own approach to teaching or learning? What possibilities excite you? What concerns do you have?

I'd love to hear your thoughts! Share your answers to these questions and list three ideas from this book that you plan to use in your own learning or teaching practice. The first 20 respondents will receive a free download of Chapter 1 from our upcoming Implementation Guide, providing practical steps to bring embodied AI into your educational environment. Comment on the website at epicleafinnovations.com or send an email directly with your responses to info@epicleafinnovations.com.

Click Below to Share your Answers!

Your insights and experiences are valuable contributions to this ongoing conversation about the future of education. By sharing your perspective, you're not just reflecting on your own practice—you're helping to shape the evolving landscape of AI-enhanced learning.

Remember, the journey of embodied AI education is just beginning. Your engagement, your questions, and your innovations will help write the next chapters in this exciting field. So, take a moment to reflect, share your thoughts, and join us in reimagining education for the AI age.

Together, let's unlock the transformative potential of embodied AI education!

About the Author

Johnna Haskell is an adventurer at heart, looking for transformative experiences in pedagogy, sports, flying, and everyday life.

Johnna has a PhD in curriculum studies/science education with extensive research in flow pedagogy of outdoor education. She completed her MSEd in educational leadership. She attained her BS degree in animal science while minoring in studio art and outdoor education.

Johnna's work concentrates on flow pedagogy and outdoor embodied experience. Johnna's broad depth of experience includes freelance wildlife photography, science educator, administrator, and university teacher for over 18 years in education before shifting her passion back to sharing her extensive perspectives on the body-mind-world.

She has explored the outdoors via canoe, kayak, skis, rock and ice climbing, backpacking, mountaineering, and paragliding. Her pursuits have included climbing ascents of Denali (20,320 feet) and an attempt to paraglide off the summit of Kilimanjaro (19,341 feet) for a major charity event to help with water supplies in various African communities.

Did You Enjoy Reading Embodied AI Education?

Thank you for spending your time with me in this book. I hope you found the experience rewarding.

If you enjoyed reading *Embodied AI Education: Unlocking Human Potential Through Enactive Learning* and found it helpful, would you mind doing me a small favor? Would you be willing to leave a short review for the book on Amazon? A sentence or two about something you liked would mean the world to me.

Potential readers would value your words and make a greater impact on people finding the book. Also, I plan to write the implementation guidebook next. Please look for new releases by following me on my Author page. If you would like to be notified when these books are released and take advantage of reduced prices, be sure to join the mailing list on my web page: epicleafinnovations.com

Join the Mailing list Below!   Please leave your review!

Please don't hesitate to share your tips or techniques or reach out to me if you have any questions at info@epicleafinnovations.com

Johnna Haskell

References

Abram, D. (1996). *The spell of the sensuous: Perception and language in a more-than-human world*. Vintage Books.

AI-Scholar. (2023). *AttenFace: Real-time attendance verification using facial recognition*. https://ai-scholar.tech

Bauld, A. (2021, May 19). *Learning in digital worlds*. Harvard Graduate School of Education. https://www.gse.harvard.edu/news/21/05/learning-digital-worlds

Capra, F. (1996). *The web of life: A new scientific understanding of living systems*. Anchor Books.

Carnegie Mellon University. (2020). *New AI enables teachers to rapidly develop intelligent tutoring systems*. https://www.cmu.edu/news/stories/archives/2020/may/intelligent-tutors.html

Chien, Y.-C., Su, Y.-N., Wu, T.-T., & Huang, Y.-M. (2019). *Enhancing students' botanical learning by using augmented reality*. Universal Access in the Information Society, 18(2), 231-241. https://doi.org/10.1007/s10209-017-0590-4

Cooper, S., Khatib, F., Treuille, A., Barbero, J., Lee, J., Beenen, M., Leaver-Fay, A., Baker, D., Popović, Z., & Foldit Players. (2010). Predicting protein structures with a multiplayer online game. *Nature, 466*(7307), 756-760.

Coursera. (2023). *Generative AI in practice: How Coursera built and implemented its foundational GenAI learning strategy*. Retrieved from https://www.coursera.org/enterprise/articles/how-coursera-built-genai-learning-strategy-cm

Csikszentmihalyi, M. (1975). *Beyond boredom and anxiety: Experiencing flow in work and play.* San Francisco: Jossey-Bass.

Csikszentmihalyi, M. (1990). *Flow: The psychology of optimal experience.* Harper & Row.

Degreed. (2023). *Artificial intelligence in learning and development.* https://explore.degreed.com/artificial-intelligence/

Dewey, J. (1938). Experience and education. Kappa Delta Pi.

Digital Promise. (2023). Micro-credentials. https://digitalpromise.org/initiative/educator-micro-credentials/

Edutopia. (2021). *AI professional development helps teachers with tech integration.* https://www.edutopia.org/article/ai-professional-development-helps-teachers-tech-integration

Edutopia. (2023). *AI professional development helps teachers with tech integration.* https://www.edutopia.org/article/ai-professional-development-helps-teachers-tech-integration

edX. (2023). *edX debuts two AI-powered learning assistants built on ChatGPT.* https://press.edx.org/edx-debuts-two-ai-powered-learning-assistants-built-on-chatgpt

Eneza Education. (2024). *Transforming education: The impact and future of AI in EdTech.* https://www.enezaeducation.com/2024/02/17/transforming-education-the-impact-and-future-of-ai-in-edtech/

Fekete, M., Lehoczki, A., Major, D., Fazekas-Pongor, V., Csípő, T., Tarantini, S., Csizmadia, Z., & Varga, J. T. (2024). Exploring the influence of gut–brain axis modulation on cognitive health:

A comprehensive review of prebiotics, probiotics, and symbiotics. *Nutrients,* *16*(6), 789. https://doi.org/10.3390/nu16060789

Forbes. (2024). *The emotional intelligence imperative in the age of AI.* https://www.forbes.com/councils/forbeshumanresourcescouncil/2024/10/09/the-emotional-intelligence-imperative-in-the-age-of-ai/

Forward Pathway. (2023). *The role of AI in sustainable development and interdisciplinary education.* https://www.forwardpathway.us/the-role-of-ai-in-sustainable-development-and-interdisciplinary-education

Gallagher, S. (2005). *How the body shapes the mind.* Oxford University Press.

Goleman, D. (1995). *Emotional intelligence: Why it can matter more than IQ.* Bantam Books.

Goel, A. K., & Polepeddi, L. (2018). *Jill Watson: A virtual teaching assistant for online education.* Georgia Institute of Technology. https://repository.gatech.edu/handle/1853/59104

Graesser, A. C., D'Mello, S. K., & Pessoa, L. (2012). Meta-knowledge in tutoring. In M. J. Lawson & J. R. Kirby (Eds.), *The quality of learning: Dispositions, instruction, and mental structures.* Cambridge University Press.

Hallas, T. (2023). *Tech column: using AI in the music classroom, part 1.* Music Teacher magazine.

Hallas, T. (2024). *Tech column: using AI in the music classroom, part 2.* Music Teacher magazine.

Harvard Extension School. (2023). *Graduate certificate in artificial intelligence.*

https://extension.harvard.edu/academics/programs/graduate-certificate-in-artificial-intelligence/

Harvard Graduate School of Education. (2024). *How generative AI can support professional learning for teachers.* https://www.gse.harvard.edu/ideas/usable-knowledge/24/10/how-generative-ai-can-support-professional-learning-teachers

Hernandez, G. (2018*). AltSchool's out: Personalized learning platform exiting standalone school business.* EdSurge.

HireVue. (2024). *AI in recruiting: Ethical & effective AI hiring.* https://www.hirevue.com/ai-in-hiring

Ibáñez, M. B., & Delgado-Kloos, C. (2018). Augmented reality for STEM learning: A systematic review. *Computers & Education, 123*, 109-123. https://doi.org/10.1016/j.compedu.2018.05.002

Immordino-Yang, M. H., Darling-Hammond, L., & Krone, C. (2019). *The brain basis for integrated social, emotional, and academic development: How emotions and social relationships drive learning.* Aspen Institute.

Institute of Education Sciences, What Works Clearinghouse. (2010, August). *Cognitive tutor (IES Publication No. WWC IRALGT10).* U.S. Department of Education. https://ies.ed.gov/ncee/wwc/Docs/InterventionReports/wwc_cogtutor_083110.pdf

Long, D., & Magerko, B. (2020). *What is AI literacy? Competencies and design considerations.*

Mavrikis, M., Grawemeyer, B., Hansen, A., & Gutiérrez-Santos, S. (2019). Exploring the potential of speech recognition to support problem solving and reflection. In *Artificial Intelligence in Education* (pp. 210-222). Springer.

McKinsey & Company. (2020). *How artificial intelligence will impact K–12 teachers.* https://www.mckinsey.com/industries/education/our-insights/how-artificial-intelligence-will-impact-k-12-teachers

Merleau-Ponty, M. (2012). *Phenomenology of perception* (D. A. Landes, Trans.). Routledge. (Original work published 1945)

Ministry of Finance. (2019). *AuroraAI – Towards a human-centric society. AuroraAI development and implementation plan 2019–2023.* https://vm.fi/documents/10623/1464506/AuroraAI+development+and+implementation+plan+2019%E2%80%932023.pdf

MIT Media Lab. (2018). *AI Ethics and Governance Initiative.* https://www.media.mit.edu/groups/ethics-and-governance/overview/

Minerva University. (2023a). *Empowering responsible technology education: AI consensus initiative.* https://www.minerva.edu/blog/empowering-responsible-technology-education-ai-consensus-initiative/

Minerva University. (2023b*). Students@AI: What does it mean to be a student in the age of AI?* https://www.minerva.edu/blog/students-ai-what-does-it-mean-to-be-a-student-in-the-age-of-ai/

NVIDIA Developer. (2023). *Artificial intelligence helps grade exams 90% faster.* https://developer.nvidia.com/blog/artificial-intelligence-helps-grade-exams-90-f